REVOLUTIONS IN SCIENCE

Turing and the Universal Machine

The Making of the Modern Computer

Jon Agar

Series editor: Jon Turney

ICON BOOKS UK
TOTEM BOOKS USA

Published in the UK in 2001
by Icon Books Ltd., Grange Road,
Duxford, Cambridge CB2 4QF
E-mail: info@iconbooks.co.uk
www.iconbooks.co.uk

Published in the USA in 2001
by Totem Books
Inquiries to: Icon Books Ltd.,
Grange Road, Duxford,
Cambridge CB2 4QF, UK

Sold in the UK, Europe, South Africa
and Asia by Faber and Faber Ltd.,
3 Queen Square, London WC1N 3AU
or their agents

Distributed to the trade in the USA
by National Book Network Inc.,
4720 Boston Way, Lanham,
Maryland 20706

Distributed in the UK, Europe,
South Africa and Asia by
Macmillan Distribution Ltd.,
Houndmills, Basingstoke RG21 6XS

Distributed in Canada by
Penguin Books Canada,
10 Alcorn Avenue, Suite 300,
Toronto, Ontario M4V 3B2

Published in Australia in 2001
by Allen & Unwin Pty. Ltd.,
PO Box 8500, 83 Alexander Street,
Crows Nest, NSW 2065

ISBN 1 84046 250 7

Series editor: Jon Turney

Originating editor: Simon Flynn

Typesetting by Hands Fotoset

Printed and bound in the UK by
Cox & Wyman Ltd., Reading

Contents

Acknowledgements iv

1 Universal Machines 1
2 The Blue Pig 3
3 Prolific Piglets: Computers everywhere 6
4 A World Out of Control 11
5 Babbage: A 'Difference Engine' that made
   a difference? 14
6 The Analytical Engine 21
7 Accidental Information 26
8 Information on the Masses 31
9 The Spur of War 39
10 Zuse in Nazi Germany 41
11 American Speed 53
12 Turing in Secret Britain 63
13 Foundations Shaken 71
14 Mathematics: Truth or game? 74
15 Crisis Looms 80
16 Turing and the Decision Problem 85
17 Government Codes 101
18 The Computer 113
19 Minding the Gap: Many universal machines 120
20 Cold War Minds 135
21 Materialisation 140
Further Reading 151

Acknowledgements

I would like to thank Jon Turney and Bat for reading my manuscript and suggesting improvements. (Remaining errors, of course, should be blamed on me – or on the universal machine that helped me write this book.)

1 Universal Machines

Take out a Swiss Army knife and have a good look at it. I have one here. It has the full range of gizmos and attachments. There is a pair of scissors, a retractable pen, a ruler, a magnetic Phillips screwdriver, some tweezers, a small blade and an emergency blade. There is even a 'cuticle pusher' and a nail file, essential for any well-manicured soldier. Nothing to get stones out of horses' hooves, but very handy nevertheless.

Swiss Army knives are versatile machines: they can be put to many different uses. Other machines are far more restricted. A lawnmower, for example, can mow lawns, but not much else. It has been designed for a specific purpose, and the function of each part of it follows. The handle is there so that it can be pushed by an adult human. The engine will power the blades, which would be exhausting to turn by hand. The blades are set so that grass is cut to an inch off the ground, the height we like lawns to be. While the lawnmower can be put to other purposes – propping open a door, perhaps – it will usually not be very effective. No one tries to fly the Atlantic on a lawnmower. Flying requires different kinds of special-purpose machines.

Some devices are more versatile because they are simple. A sharpened stick, for example, can be used as a lever, or to cook a kebab, or to knit a sweater. Indeed, more uses can probably be found for a simple sharpened stick than for a Victorinox Pocket Size MiniChamp II – my top-of-the-range Swiss knife. Yet, despite their varying versatility, Swiss Army knives, lawnmowers and sharpened sticks are all a similar sort of machine. Even the knife and the stick are, in the end, *special-purpose machines*, and are radically different to an astonishing device built for the first time in the middle of the last century: a machine of *universal* application.

2   The Blue Pig

An early example could be found in Manchester in 1951. It filled a room, and broke down regularly. A team of engineers tended it, replacing the valves – or vacuum tubes – as they blew. They called it the 'Blue Pig'. If you had £150,000 you could buy one of these machines for yourself, although there would be a queue of military establishments and scientific laboratories ahead of you. Three years earlier, the first ever machine of this type had been built a hundred yards away. That one was an experiment, rows of electronic tubes and a tangle of gutta-percha-covered wires filling what resembled a set of bookshelves. The 1951 model gleamed – the valves hidden in banks of metal cupboards, a shiny central console with rows of switches and lights.

Late in the year, the Blue Pig had some visitors. They were from a children's radio programme, and had come to hear the Pig sing. The engineers prepared the machine, and, after a moment's hesitation, a gratingly harsh but stately National Anthem blared forth. The radio presenter was delighted. The patriotic hymn was followed by 'Baa Baa Black Sheep' and finally the dancehall jazz of 'In the Mood'. The Blue Pig had trouble with the last tune: it improvised some notes of its own and then fell into silence. The machine, concluded the radio presenter, was not, after all, in the mood.

With the visitors gone, the engineers returned to another task, but with the same machine. The Pig could produce poetry, doggerel love letters. Here's an example:

Darling Sweetheart,
You are my fellow feeling. My affection curiously
clings to your passionate wish. My liking yearns
to your heart. You are my wistful sympathy: my
tender liking.
    Yours beautifully,
    M.U.C.

The Blue Pig could do mathematics too. Much faster than any human mathematician, it made calculation after calculation. What it searched for were moments when a certain function – the Riemann Zeta function – took the value of zero. It was something of a fishing expedition, but if they were lucky and found an unexpected zero, then a famous mathematical hypothesis would be proven wrong. Despite the Pig's all-night efforts, none was found. This was a particular disappointment to a middle-aged man of awkward manner, who had achieved early fame proving another hypothesis wrong – and at the very same moment had come up with the idea now expressed in massive material form by the Blue Pig. This man was Alan Turing, and the renaissance Pig – one machine producing music, poetry and mathematics – was MUC: the Manchester University Computer.

3   Prolific Piglets:
Computers everywhere

Computers nowadays look nothing like the Blue Pig. But the machine that sits on your desk shares the same ability as its predecessor from half a century ago: it is a universal machine. I use mine to write books, send e-mails, play music and add numbers. Without it I'd need a typewriter, a pen, paper, envelopes, a CD player and a calculator. If I stuck all these special-purpose objects together I'd have something akin to a very strange Swiss Army knife, but it wouldn't be a computer. A computer can switch between these different jobs, while apparently remaining the same machine. It is a different sort of device from the special-purpose machines.

What makes the difference is the fact that a computer has two parts: a part which does stuff, and a part which has a list of instructions of what stuff to do next. It is the fact that we can change the list of instructions that makes the computer a different sort of device from a typewriter or a calculator. In other words, the computer can store and run a program.

There is another sense in which computers in the twenty-first century are universal machines: they seem to be everywhere. Not only do they sit atop millions of desks in every city, but they also come in many sizes and shapes. Laboratories such as CERN on the Swiss–French border or Los Alamos in New Mexico, where thousands of scientists congregate, depend on massive, powerful computers, each of which might cost as much as a new hospital. These machines are bigger and pricier than the Blue Pig. However, the drop in the cost of computing in the twentieth century was something unprecedented in the history of technology. As one chief executive put it: if the performance and cost of the car had followed that of the computer, we would be driving to Mars on a thimbleful of gasoline. The reason for this remarkable transformation is miniaturisation: by the 1970s a whole computer could be put on one small chip of silicon, a microprocessor. That's a Blue Pig sitting on something the size of a postage stamp.

As a result of this miniaturisation, computing

has become too cheap to meter. Take, for example, a washing machine. By the late nineteenth century there was already a range of special-purpose machines to help clean clothes, including two important ones. First, a big vat in which clothes could be soaked, soaped and stirred. Using the vat, you got clean but wet clothes. Second, a mangle, which consisted of two rollers attached to a handle. You fed the clothes between the rollers and cranked the handle to squeeze out the water, and the end result was clean and almost dry clothes. It was back-breaking work for a human: stirring, lifting and cranking. Then the washing machine added another device – an electric motor – and changed how the work was done. Think carefully about what has happened. Now you just put the clothes into a washer-dryer, press a button, wait, and – ideally – take out clean, dry clothes. To do this, the machine has to have parts which clean and parts which dry. However, in any piece of work something else must be clear too: knowledge of the order in which to perform the stages of the work. Or, in other words, any job can be expressed as a list of instructions. The list that a washing machine has to follow is not very long, nor very complicated. However, in a modern washing machine this list is programmed into a microprocessor. We know that a computer can follow any list of instructions – as long as they are put in a form that it can understand. It is an indication of just how cheap computers have become that we now put them in washing machines.

The microprocessor of a washing machine governs the order of rinses and spins. The miniature embedded computers found in modern cars monitor the engine, keep track of fuel consumption, and control the information displayed on the dashboard. Even the roads that the car travels upon are getting smarter: microprocessors control the traffic lights, the hazard warning signs, the speed cameras, and the electronic panels which flash up the number of free parking spaces in the city centre. Electronic information is constantly circulating in the modern world, a flow enabled by millions of miniature computers. The universal machine is not just a device of multi-purpose application, but has become ubiquitous, if often unnoticed. In the West, in the year of the Y2K scare, the average distance from every human to a computer was a matter of a handful of metres.

Computers present a strange case in the history

of technology. They are machines of apparently limitless applicability, yet they are also the drudges of the modern world. Numbering millions, they have a typical working day made up of repetition, repetition, repetition. How can the invention of this remarkable device be explained? The question is the same as asking: what sort of society would ever need such a thing?

4 A World Out of Control

The world we inhabit is one ordered by large organisations: big business corporations and government bureaucracies. What these two have in common are managers: while the work at the bottom of the pyramid is often repetitive and specialised, at the top the skills required are more strategic, more general. This two-tier modern world of general- and special-purpose humans was built in the nineteenth century as a counter-revolution. Over the previous century, transformations in industry had produced a landscape that contemporaries viewed with anxiety: sprawling cities inhabited by uprooted discontents, spasms of unrest, cholera and chaos. Industrial society was pandemonium, hellish and running amok. The answer to the world out of control was organisation: the corporate business and the big government department. Yet there was something else going on too.

As corporations and bureaucracies have grown, so they have speeded up and been mechanised. From the late nineteenth century, organisations have crystallised from within. In the twentieth century this mechanisation gathered pace, and now your tax calculations or insurance claims pass through silicon semiconductors as well as clerical hands. In answer to the question of what sort of society would ever need a general-purpose machine, we have a clue to where and when to start looking: we know that in the nineteenth century there emerged organisations that embodied in the manager and clerk the general-purpose/special-purpose split. But we have also generated another question: we have to explain mechanisation. So, this story of the universal, general-purpose machine will start in the nineteenth century, with Charles Babbage, since no one understood the changing nature of industrial society and mechanical control better than he.

Figure 1: Charles Babbage (1791–1871), Lucasian
Professor of Mathematics, Cambridge. Stipple engraving
after John Linnell, 1832. Published by Colnaghi, Pall
Mall East, London, 1 January 1833. (Source: Science
Museum.)

5   Babbage: A 'Difference Engine' that made a difference?

In 1819, Charles Babbage set out for France with his friend, John Herschel. Both were young men in their late twenties, scions of an English scientific world of which they were increasingly critical. John was the son of the German émigré astronomer and celebrated discoverer of the planet Uranus, William Herschel. Babbage had not had such an easy debut. He had taught himself mathematics, and entered Cambridge University at the age of 19, only to find himself already ahead of his cohorts. The University was largely a finishing school for gentlemen, and its mathematics lagged seriously behind the strange country Babbage and Herschel set out to tour. Four years previously, Napoleon's army of revolutionary France had been defeated at Waterloo, and only now could contact be made with the old enemy.

Babbage was a polymath: not only was he a

mathematician, but he also wrote on economics, politics, astronomy and engineering. He advised insurance companies and had lectured at the Royal Institution. However, time and time again he encountered the same problem, one that seems trivial to us but would dominate his life: mistakes in mathematical tables. Students in the twenty-first century can pass through all their years of study without ever seeing a mathematical table, yet in Babbage's time they were a matter of life and death. Babbage and Herschel's England was at the centre of two great historical tidal waves: industrialisation and empire, and both created an insatiable demand for mathematical tables.

Take the Empire first. British imperial power was only as strong as its navy, a massive technological system of ships, sailors, ports and weapons, linking the dockyards of Chatham, Portsmouth and Devonport to the colonies of India, the Cape, Australia, New Zealand and Canada. The Royal Navy protected the burgeoning trade of the merchant ships, and thereby the wealth and power of Britain. Yet navigation was treacherous, and knowledge of exact locations was needed by every ship to avoid reefs and rocks. To turn measurements of the positions of stars into such knowledge required tables. A mistake in a table could lead directly to shipwreck, with loss of life and – felt as deeply in London counting houses – loss of cargo. Fortunes depended upon accurate calculation.

The new markets created by Empire fed the astonishing transformation underway in the home country in the production of new things to sell. Cotton, for example, was bought raw in bulk, transported across the oceans, unloaded at the bustling Liverpool docks, and distributed to a number of unsettling and even frightening new buildings. In these 'manufactories' – we now shorten the term – hundreds of men, women and children might work under one roof. The rhythm of their work was not of their own choosing, since they were mere attendants to the machines that wove the cotton into cloth; and these machines were driven not by human or animal muscle but by steam engines. Their work was also repetition, repetition, repetition: a complicated multi-stage job such as producing cloth from raw cotton had been broken down into simple components, and the factory strung together the components, organising the order of the factory hands operating their single-purpose machines.

The workers had emigrated from the countryside and created new cities: Manchester, Leeds, Birmingham. Torn from intimate village life, they now lived in cities of strangers – and cities of death (the industrial city killed many more people than were born there). Babbage was well aware that industry too demanded tables. Engineering, rapidly passing from craft knowledge to science, required such grids of numbers to assist calculation. Even the industrial dead found their place in the tables. The fear of dying was intensified in the industrial cities by the extra anxiety of dying a pauper and not being able to afford a respectable grave. Life insurance had blossomed on the back of this industrial fear, and once again reams of calculation were needed. One of Babbage's jobs had been to assist the calculation of new mortality tables. Both industrialisation and Empire demanded the production of vast numbers of accurate tables, and inaccurate tables cost lives and money.

Babbage considered the problem of tables while travelling in France. Babbage and Herschel's tour carried them to the pre-eminent organiser of revolutionary French mathematics, Gaspard Clair François Marie Riche de Prony, lecturer at the engineering school – the École Polytechnique, director of the École des Ponts et Chaussées, and author of the multi-volume *Nouvelle Architecture Hydraulique*. The English visitors witnessed Gaspard de Prony's latest organisation. He had broken down the work of calculating tables into simple parts. Instead of skilled mathematicians slowly calculating numbers, the labour had been so divided that the barely numerate could be employed in teams to work through the calculation, using a procedure called the 'method of differences'. In fact, so simple was the mathematics that resulted from this division of intellectual labour that even hairdressers, left idle after ornate wigs had fallen from fashion with the execution of Louis XVI, could be assigned to the task. Babbage was enthralled. Few in Britain could match his knowledge of manufacturing, especially the division of labour that factories exhibited; yet here was an application that was new to him. The table problem was threefold: numbers had to be produced quickly, calculated accurately and printed correctly. De Prony's system was fast – faster than any individual mathematician – and produced numbers in bulk, but it was still inaccurate, a fact which Babbage blamed on the human operators still very much part of the process.

Back in Britain, Babbage continued to think about tables. While France was pre-eminent in organisation and analysis (the science of breaking down things into component parts), England had steam manufacture. Babbage was in a unique position of possessing considerable knowledge of both, and, in a decisive moment, he put the two together. While working on an astronomical table with John Herschel in 1820, Babbage later recalled:

> [I]*t was suggested by one of us, in a manner which certainly at the time was not altogether serious, that it would be extremely convenient if a steam engine could be contrived to execute calculations for us; to which it was replied that such a thing was quite possible.*

Powered by steam, machinery would automate de Prony's method of differences, printing out tables as an end product. With the human role limited to setting up and starting the automatic calculating machine, the tables would be produced quickly and as accurately as any mechanical process. Babbage called this project the Difference Engine.

For 12 years Babbage tried to build a Difference Engine. He received a substantial grant of £1,500 from the government, but only portions of the Engine were ever completed. Partly, this failure was due to runaway costs (like many big science projects, it exceeded its budget, eating up more than £30,000, much of it Babbage's own money). Clashes with his chief engineer, Joseph Clement, were a testimony to the technical difficulty of the precision engineering needed to build the Engine. But the project was well within the capabilities of early nineteenth-century advanced engineering techniques. The greatest handicap to the Difference Engine was that its designer became distracted by a second project, one of grand sweep and consequence for our story.

6   The Analytical Engine

Rather than automate a specific technique – the method of differences – Babbage had thought of a design for an automatic general-purpose calculating machine, which he called the Analytical Engine. Again, the inspiration came from the factory organisation that Babbage knew so well. What if, he reasoned, he could design a mill for manufacturing numbers – a calculating section that could carry out mechanically all the basic rules of arithmetic? Separate from the mathematical mill would be a store – just as a cotton mill had a separate store for the cotton bales – for the numbers that were to be used in the calculation. All that was needed was a third component: a means of controlling the Engine, the instructions that would tell the machine which numbers to bring from the store to the mill, and what was to be done with them once there. A human could, of course, play this controlling role. But Babbage sought to automate the entire process, and searched his experience of industry for inspiration. He hit upon the Jacquard Loom: a cloth-making machine which carried the pattern to be woven in the form of holes punched in cards. Such a loom could weave a new design just by changing the card program. The Analytical Engine could likewise shift flexibly between calculations: at one moment it could imitate a Difference Engine if the cards held instructions for the method of differences, but the next moment the cards could be changed and the Engine would churn out mortality tables. As Ada Augusta, Countess of Lovelace, Babbage's friend and colleague, wrote in her account of operating the Engine (the most detailed we have):

[W]e may say most aptly that the Analytical Engine weaves algebraical patterns just as the Jacquard Loom weaves flowers and leaves.

Babbage's Analytical Engine is sometimes claimed as the first computer – which it both was and wasn't. It wasn't, not only because it was never completed, but also for a subtler reason: the computer was an

Figure 2: Henry Babbage's Analytical Engine Mill, 1910. Charles Babbage did not complete his Analytical Engine. However, nearly four decades after Charles's death, his youngest son, Henry Provost Babbage, built a hand-operated printing calculator based on plans for the Analytical Engine's mill. (Source: Science Museum.)

object of the twentieth century and had meaning in twentieth-century terms (as it will have different meanings by the middle of the twenty-first).

Babbage's machine was rooted in the nineteenth century: everything it did spoke to the concerns of the day – witness its terminology of stores, mills, looms and engines. Babbage's techniques of calculation and control depended on different media because he, like his contemporaries, saw the two processes as distinct – the former used decimal, brass cogs, and the latter Jacquard Loom cards. (We will see later that the alternative choice, to blur the distinction between instructions and data, was a profound one.) Such nuances matter because they tell us about what machines really meant for the societies that built them. Yet there is also a continuity between Babbage's time and ours which justifies starting with him: it was an Engine proposed because of crises of industrialisation; it was a machine that automated the mental process of mathematics because those calculations were essential to an industrial world that was generating more information than could be handled by humans alone. And we are also in a world where industrial processes multiply the problems of handling information, now perhaps a million-fold from Babbage's time. Since we are confronted by the same problem as Babbage encountered, albeit one greatly magnified, our solutions share characteristics. One is the drive to automate, to speed up the process of thought through extensive mechanisation. So, while I would hesitate to call the Analytical Engine a computer, an enquiry as to why they are similar leads us to ask what characteristics of the surrounding society were shared by the two.

But something is missing in the explanation I've given so far. I have argued that industrial societies threw up thorny problems of mass calculation. Machines along the special-purpose lines of the Difference Engine would have helped here. But what would explain the proposal for a *general-purpose* machine such as the remarkable Analytical Engine, a device of immense flexibility, of almost universal application? This is a question I shall return to because it sheds light on Turing and the universal computer. But before it can be answered, we need to understand matters of big business, government and war – and railway disasters.

7   Accidental Information

In September 1830, Babbage had travelled to Manchester to witness the opening of the Manchester and Liverpool Railway, the first passenger line in the world. The day before, at the Exchange – the commercial heart of the industrial city – he had been introduced to William Huskisson, a member of parliament and previously Colonial Secretary in Prime Minister the Duke of Wellington's administration. On the day of the opening, Babbage had joined the crowded train as it made leisurely progress from Liverpool to Manchester. Then, as he recalled in his autobiography,

> *the whole of our trains came to a standstill without any ostensible cause. After some time spent in various conjectures, a single engine almost flew past us on the other line of the rail, drawing with it the ornamental car which the*

Duke of Wellington and other officials had so recently occupied. Instead of its former numerous company it appeared to convey only two, or at most three persons; but the rapidity of its flight prevented any close observation of the passengers. A certain amount of alarm now began to pervade the trains, and various conjectures were afloat of some serious accident.

In fact, Huskisson was dead, run over and fatally injured by Stephenson's Rocket. Industry out of control had claimed a famous victim.

Although Babbage could not have realised, there was to be a direct link between railway accidents and the modern world of information flow. In Britain, anxieties over political control focused on a discontented working class, crowding the industrial cities – at that very moment, a massive mutinous crowd had gathered to greet Wellington with revolutionary tricolours, and the chief officer of Manchester had insisted that the procession must continue, despite the accident, because otherwise 'he could not be answerable to the safety of the town' – the Rocket must arrive or there would be riots. In the United States, the anxiety over control was a problem of space, not people. 'From sea to shining sea', the enormity of the new continent created crises of organisation.

On 5 October 1841, two passenger trains on the Western Railroad, which took a long and mountainous route between Worcester and Albany, collided head-on, killing two. A public outcry and inquest followed. Chastened, the owners of the Western line proposed a radical solution. Technological systems, such as the railroad, were now, they argued, too large for direct control by single individuals. What was needed, drawing an analogy directly from army organisation, was a line of command and the creation of a hierarchy of managers at local, regional and headquarters levels. Such a structure would prevent accidents because the managers could enforce the collection and processing of information. Strict rules about when and how trains should run along the tracks, and how to report such information, were introduced. The prime tool of the new managerial hierarchy was information. Beginning with the railroads, this new managerial style – the corporation – spread through American industry, each business being forced, as soon as a certain scale was reached, to tackle such crises of control. Corporations have dominated the economies of the Western world ever since, and they rely on techniques to handle information.

So the scale of the American continent caused the story of communication and information machines to shift emphasis. Technologies that were invented in Europe were pushed to extremes in the United States, and were transformed in the process. The railroad was one example, and another was the telegraph. The two spread hand-in-hand, with telegraph lines alongside the train tracks. Partly this co-development was a matter of convenience. After all, both linked towns and cities between which goods and information travelled. But, also, the instantaneous communication that the telegraph provided was precisely what the railroad managers required: a message flashed down the line could warn an approaching train of a derailment, as well as pass on everyday information about the state of the business. Railroad corporations and telegraphs are the perfect exemplar of a general pattern: industrial technology would run out of control without a means of processing prodigious quantities of information.

Now the managers of the new corporations

might decide what information should be collected and where it should be sent, but the actual processing of nineteenth-century information was the work of armies of human clerks. The central Railway Clearing House in Britain, for example, by 1876 employed 1,440 clerks under one roof. The British railway system then, as now, consisted of a hotchpotch of private companies. But travellers often wanted to make complex journeys, joining several trains under different ownership. The question was: how would the money received for the ticket be fairly divided between the companies? This is what the Clearing House achieved, using its thousands of clerks to process masses of data, aided only by simple information technologies: standardised forms, ledger books, pigeonhole desks and very basic mechanical adding machines. The new big businesses of the nineteenth century were often competing for such small change, because with their economies of scale the small change would add up to big profits. In this way, mass industries – mass production, mass distribution and mass consumption – led to mass information processing.

8   Information on the Masses

Yet in many ways the challenges encountered by the corporations and big businesses were already familiar to a much older organisation that defined itself by its ability to control a geographical area: the state, with its government and army. And while the corporations employed many innovations – typewriters, document copying machines, the vertical filing system – in their desperate attempts to manage the information they generated, one of the most significant turning points in the history of information technology was a response to a public rather than private crisis of control.

All governments require knowledge of the population in order to act. In modern times such knowledge has been acquired through the organisation of regular censuses. The first census of the population of the United States took place in 1790, not long, of course, after the revolution. They have been held every ten years ever since. Through the nineteenth century, two factors made the census increasingly difficult: the population was growing, so there were more people to survey; and, against familiar resistance, the federal government was taking on more responsibilities and therefore needed a greater range of information to be collected. The pressure began to overwhelm the hand-collected, hand-processed census. The 1880 Census took seven years to work through – by the time the information was put into a useful form it was almost time for the next census. The government was acting on nearly decade-old knowledge of the population, at a time when immigration was expected to double the figure. Such knowledge was nearly useless. The 1890 Census was expected to be even more complex: seeking information on 235 topics per person compared with 215 in 1880, and a mere five in 1870.

The Director of the Census, the British-born Robert P. Porter, announced a competition for schemes to speed up tabulation. Three made it as far as a trial. Two of the entries were variants on the old manual process, but the third, from a mechanical engineer named Herman Hollerith, offered a more technical solution, based on a system to which he owned patents. In Hollerith's system the information collected would be stored in the form of holes punched into cards of standardised size, one card per person. For example, on one section of the card there was marked an 'M' and an 'F'. If the census data described a woman, then a hole punched through the 'F' (for 'Female') would record this information. Note that this was a *binary* form of storing information: the hole was either there ('1'), or it was not ('0'). Although it would be time-consuming to punch a card for every adult in the country, once completed, the advantage of Hollerith's system became apparent: a machine could be used to accept a stack of cards, and run at high speed through the stack sensing the presence or absence of holes. If the government needed to know the number of men in the United States, the job was reduced to feeding the whole stack into the machine and mechanically sorting out all the cards with a hole where the 'M' should be, and then counting them, again mechanically and at high speed.

Hollerith's punched-card system won the competition and was applied to the 1890 United States Census. A glance at the 1890 Census card reveals the kind of information that the federal government was interested in, although a certain amount of decoding is necessary. One part of the card looks like this:

US	Ir	Sc
Gr	En	Wa
Sw	FC	EC
Nw	Bo	Hu
Dk	Fr	It
Ru	Ot	Un

Most of these can be guessed: 'Ir' for Irish, 'It' for Italian, 'Sw' for Swedish and so on. The concern here was machine-processable knowledge of immigration. Another, separate block of the card was as follows:

Jp	Ch	Oc	In
	Mu	Qd	
	B		
	W		

Again the identity of the data can be guessed: 'Jp' for Japanese, 'Ch' for Chinese, 'Oc' for Occidental,

'In' for Indian, 'B' and 'W' for black and white, and 'Mu' and 'Qd' for mulatto and quadroon. The issue the government was interested in here was clearly race.

With all the cards punched, Hollerith's machines churned through the stacks, completing the job in two years, a substantial improvement on the previous census. Mass data processing had been extensively mechanised. Its system had already been celebrated: the August 1890 edition of the *Scientific American* carried a front page filled with illustrations of the machines and their minders at work. The ambitious Hollerith was determined to make a business of his invention. However, censuses were held only every ten years, so he and Porter, who was now a partner in the scheme, looked around for other potential users. His gaze naturally turned to the corporations that shared with the Census the extreme data processing problems of large-scale organisations. In 1895–6 he made a business breakthrough, persuading the New York Central Railroad to employ his punched cards to process information on a day-to-day basis. Here was a regular income for his new business, the Tabulating Machine Company. Many corporations mechanised their data processing using Hollerith's system over the next few decades: railroads, insurance companies, utilities and big industrial

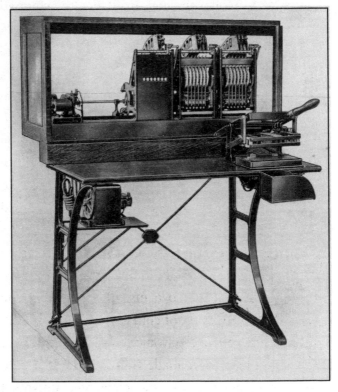

Figure 3: Hollerith Pin Box Type Tabulator. One of the first Hollerith machines marketed in Britain, in use from 1904–7. (Source: National Archive for the History of Computing.)

manufacturers. The company also sold the system to government departments, and even licensed the machines for use in censuses abroad, including one in Russia.

In 1911, Hollerith, now in ill health, stepped down from control of the company. After a merger with a cash-till company, the business was renamed and began to build up a formidable talent in sales. The company was renamed for a second time in 1924 as International Business Machines, or IBM. In competition with rivals, such as Remington Rand, IBM aggressively pushed machines into thousands of offices across the United States. At the heart of many businesses was a punched-card data processing installation, and these businesses had been reorganised around these machines so that best use could be made of them. This reorganisation will prove important later in the story. IBM today are, of course, synonymous with the manufacture of electronic computers. But there was still a leap to be made from mechanical data processing machines to the electronic universal machine, a leap spurred by war.

Figure 4: Hand-punch operator, 1950s. Punched-card systems remained in common use for over half a century. Much of the data had to be punched onto cards by hand before machinery could sort and extract the information required. (Source: National Archive for the History of Computing.)

9 The Spur of War

The modern world is in thrall to speed – and speed gives us the language to describe what's modern: acceleration, change, progress. And technology, of course, is central to this image. Think of the aeroplane, the speedboat, the motorbike, the rocket or the fast car. If we can understand how technologies change, then we are a long way to getting a grip on what makes the modern world modern. One of the greatest forces shaping technologies is warfare, and the twentieth century – the century of the computer – was ravaged by war like none before it. Without the six-year conflict at the heart of the century, the Second World War (1939–45), the history of many technologies would look radically different. Put crudely, war places great value on speed, and so encourages the development of technologies that will deliver a weapon faster, or accelerate a battleship, or rapidly redirect an anti-aircraft gun. The speed of technologies decides between the quick and the dead.

Many wartime transformations in technology involved speeding up the movement of matter. Aircraft, for example, were means of moving soldiers or bombs fast. In 1939, aircraft were powered by motor-driven propellers, and the fastest could move at 300 miles per hour. By 1945, the pressure of war had spurred the development of radically new forms: the jet aeroplane and the rocket. The German V-2 'Vengeance Weapon', launched at London, Antwerp, Brussels, Liege and Paris in the closing year of the war, used liquid oxygen and alcohol as propellants, and reached 3,500 miles per hour. That's matter moving very fast indeed.

But some of the most important ways that the Second World War transformed technology concerned moving around information rather than matter. In three countries – Germany, the United States and Great Britain – this effect was to have important consequences for the history of computing.

10   Zuse in Nazi Germany

Konrad Zuse was born in 1910 and educated in the capital of Germany, Berlin. When he entered the Berlin-Charlottenburg university in 1927, where he received one of the best engineering educations the world had to offer, his country was a republic, although an unstable one. The political state of Germany when he graduated in 1935 was utterly different: a totalitarian dictatorship under the National Socialist Party and its Führer, Adolf Hitler. Rearmament – the rebuilding of the German war machine – was underway, and Zuse soon found employment at the Henschel aircraft company, where his skills as an engineer were in demand. His work was not on engines, but on an equally important aspect of aircraft design: ensuring that the airframe of the fuselage and wings was not shaken to pieces under the intense stress of flying.

The problem of building stable airframe

structures was easy to describe mathematically: it could be expressed as a set of simultaneous equations, or, in other words, as a collection of simple numerical relationships between unknowns. An example of a simultaneous equation is as follows: supposing you know that two apples and one banana cost four Reichsmarks and one apple and four bananas cost nine Reichsmarks, then how much is an apple and how much is a banana? Get a pen and paper and work it out – and time how long it takes.

(Pause.)

How long did it take? Perhaps less than a minute? In other words, a simultaneous equation with two unknowns (the prices of apples and bananas) can be solved quite quickly with no more assistance than symbols written on a piece of paper. Now imagine a simultaneous equation with four unknowns. Say you knew that one apple, three bananas, 18 carrots and two damsons cost 15 Reichsmarks, and you had three other similar relations. Four equations and four unknowns (prices of apples, bananas, carrots and damsons), so – unless you're unlucky, mathematically – it should be solvable. And indeed it is. But when you sit down and work it out, it doesn't take merely twice as long to solve equations with twice the number of unknowns. It won't take two minutes, but longer. And this problem gets worse the more equations you have.

Zuse's airframe designs threw up simultaneous equations with 30 unknowns or more. The tension in each strut in the frame contributed an unknown, and 1930s airframes were a wickerwork of such struts. Thirty unknowns would take a team of mathematicians, each with pen and paper and sharing the work out between them (like de Prony's hairdressers), not minutes, not hours, but weeks to solve. Here was a major problem: the Henschel company was coming under considerable pressure from above to speed up the production of new military aircraft, yet the design of these machines was in danger of being held up, just because of slow mathematics. Zuse thought hard about the problem. The mathematicians were not doing anything complicated; they were mostly following simple rules: take one equation, multiply all the terms by a number, write down the new equation somewhere else on the page, find the next equation, subtract one equation from another, and so on. All repetitive stuff, and when something is repetitive – especially when explicit rules can be written about what to do next – then the work can be mechanised. Mechanical calculators had been around for centuries: Zuse's fellow countryman William Schickard had devised a machine in the 1620s that could add numbers together, and had sent drawings to the famous astronomer Johannes Kepler illustrating how it worked. We have also seen how Babbage had designed more sophisticated machines such as the Difference Engine. Indeed, mechanical calculators were in widespread use by the end of the nineteenth century, helping keep track of the accounts of the big corporations. So Zuse knew that building machines to assist calculation was not a problem, especially for simple repetitive tasks such as solving simultaneous equations. Machines such as the Difference Engine also mechanised a second part of the work of solving equations: storing the numbers (for example, of apples and bananas in the case above) that were needed whilst calculating. In other words, mechanical 'memories' were not a problem, either.

But Zuse thought about the process of solving equations further, and realised that there was a further part that could also be mechanised: the explicit instructions that the mathematicians were following. He called this the 'plan' of the calculation. Was there a way of giving the plan to the machine? If there were, then the whole process could go at a speed determined by the machine alone, since the machine would not have to stop each time it needed to know what to do next. Zuse was only a young engineer, in a lowly position in a large company, so there was no way he could put his experimental idea into practice at his employer's expense. He decided to test it out for himself.

In many ways, Zuse had reinvented the wheel: the idea of a machine with a calculating part, a mechanical memory, and a mechanised list of instructions was precisely what Babbage had hit upon with the Analytical Engine. (Think of the mill, the store, and the Jacquard cards.) The English mathematician had worked for years, with variable but generous funding from government, and with the assistance of one of the most capable engineers of his day, but his Engine had never been completed. What chance did poor Zuse have? His decisive move was to simplify calculation, not from the perspective of the human but from the perspective of the machine. Babbage had designed his Engines to calculate using decimal numbers, so called because ten symbols were needed to represent a number: 0, 1, 2, 3, 4, 5, 6, 7, 8 and 9. Decimals were easy for people to grasp: the number 'thirty-nine' was 39, all very familiar. But there were other ways of representing numbers and one of the simplest was just to use two (0 and 1), the binary system. We've seen binary before: the punched cards of Jacquard and Hollerith both stored information in binary form: either a hole was there, or not. Binary is not so human friendly. 'Thirty-nine' is 100111 – which makes little sense to us at first glance. Zuse's insight was to realise that the opposite was true from the machine's perspective. The anthropocentrism of decimals meant little to it. The explicit instructions and the numbers in the memory could be translated into binary, but the biggest effect was on calculation: since the rules of multiplication were much simpler in binary form (since one times one was one, and any other multiplication was zero) then calculation could be made much simpler. More to the point, Zuse's device could be cheap, and would not need the most skilled engineer in Germany to construct.

Notice that something else had happened, too, when Zuse decided to use binary for instructions, memory and calculation: two seemingly entirely different things – instructions and numbers – were in the same format. The difference between them could potentially blur. I'll come back to this observation later.

By 1936, Zuse had a design for an equation-solver. He presented his plan to Kurt Pannke, a local manufacturer. Pannke initially thought that Zuse must be a swindler, but eventually was persuaded to put up a little cash to fund the project. With the help of friends Helmut Schreyer, Andreas Grohmann and Walther Buttmann, Zuse started to build the machine in the living room of his parents' Berlin apartment. The memory and the calculator were mechanical. The list of instructions was first punched onto paper tape, but this proved troublesome. Then his friend had a better idea. Cinema was at its height of popularity in the 1930s, and Schreyer's hobby was as a projectionist. The feeding of celluloid through a projector suggested to Schreyer a model of how the control portion of the machine could work. The binary instructions were therefore punched, not onto paper, but onto reels of discarded film. The crude version, which he called the Versuchsmodell-1 (literally 'Experimental Model-1') was completed in 1938.

A second version was almost complete when Zuse was drafted into the German Army in the first months of the war. It was nearly a year before he could get back to work on his machines. Meanwhile Schreyer, who was an electrical engineer by training and a Nazi by politics, had experimented with speeding up the V-1. One means was to use 'relays' found in great numbers in telephone exchanges: switches that could be turned on and off by passing an electric current through them. Since relays were either on or off, they made natural binary components, and, since they were controlled by electricity, they could be fast – this aspect had saved the telephone exchanges when they had grown so large that human operators struggled to connect lines fast enough. Zuse and Schreyer took their idea for a fast relay machine, as well as their experimental versions, to the Deutsche Versuchsanstalt für Luftfaht (DVL), the national aeronautical research laboratory. The DVL was all too familiar with the problem of many-variable simultaneous equations, a problem that was developing into a crisis as the competition of war called for more and more new aeroplane designs, and money was allocated to turn Zuse's home machine into a fully engineered programmable calculator.

The Z-3 was completed in 1941 as a generally programmable, binary calculator that deployed floating-point arithmetic, with over 2,500 relays clattering away inside. It cost 25,000 Reichsmarks (roughly $6,500 of 1941 value), assisted the design of military aircraft, and remained in Zuse's house until being destroyed in an air raid in 1944. The electrical engineer, Schreyer, was even more ambitious: why not replace the relays with an even faster technology: electronics, which were at the centre of a whole industry – radio – that had grown in the first half of the twentieth century. Relays were always going to be limited by the fact that to switch one from on to off, a physical piece of metal had to be moved. Inertia has to be overcome as it is moved, and this inertia was considerable compared to that of an electron. In a vacuum tube – or valve, depending on which side of the Atlantic you are from – electrons were accelerated through a vacuum. If current flows, the switch is 'on'; if not, the switch is 'off'. Valves are not much different from an electric lightbulb; both work by passing an electric current through a thin wire or 'filament'. In a lightbulb, the current merely serves to heat up the filament to the point when it glows brightly. In a valve, there are at least three filaments. Current can pass through one, boiling off electrons as it does. The middle wire can either attract or repel the electrons. If the electrons are attracted, they leap across the vacuum to the third wire, and current flows. (If the electrons are attracted by a particularly strong positive middle wire, the current is amplified, a phenomenon that made radio and television possible.) If the electrons are repelled, the current stops. The state of the middle wire – positive or negative – therefore acts like a switch. So a valve is like a particularly sophisticated lightbulb, one that can be used to amplify current or imitate a relay. Vacuum tubes could therefore also be used as binary devices, and could outpace the relatively lethargic relay by a factor of thousands. Schreyer had contacts within the Nazi party, but despite gaining the opportunity to present his scheme for an electronic programmable machine to the German High Command, he was thwarted by an instruction that came from the Führer himself. Hitler had decreed, as the war turned against Germany at Stalingrad, that only new technologies that would directly assist the war effort would henceforth be funded. The terror weapons, such as the V-2 rocket, were approved; Schreyer's project was not. A different decision might have seen an electronic valve computer built in Nazi Germany – a universal machine for the Third Reich. Such were the contingencies of war: at one moment fostering fast development, at another halting development altogether.

Zuse and Schreyer have always been relatively obscure figures in the history of computing. There are at least a couple of reasons for this. Their work was dispersed in the whirlwind of destruction in the closing year of the European war. The Nazi Schreyer left soon after for Rio de Janeiro, and Zuse moved to Switzerland, taking parts of his Z-4 machine with him. He went on to design further computers, and in doing so devised the first special-purpose programming language, called 'Plankalkul'. But continental Europe was in ruins, and high technology was not the first priority in the early years of reconstruction. Another reason why Zuse's name is not well known is that fame needs institutional investment. The celebrated figures of twentieth-century computing – with Turing being a notable exception – are often associated with powerful corporate interests. To find an example we now cross the Atlantic.

11   American Speed

Howard H. Aiken was a formidable applied mathematician, completing a thesis at the prestigious Harvard University. Like Zuse, he had frequently been confronted by equations that required enormous quantities of time and resources to solve, and began to investigate mechanical means of tackling them. By the 1930s, several laboratories in the world were deploying Hollerith-type systems, not to process business data, but to solve complex scientific calculations. At these laboratories, humans still did the organising or 'control' work – deciding which cards to feed into the machine next, and what the machine should do with them. Aiken had seen something similar at the Thomas J. Watson Astronomical Computing Bureau at Columbia University, and continued the interest. Unlike Zuse, Aiken was something of a historian of his subject, and knew of Babbage's nineteenth-century

Engines. He reasoned that punched-card machines could take the place of the brass calculating mill. Punched cards could also store the numbers required. And holes punched into paper tape could be used to hold the instructions, replacing the human 'controls' and telling the calculating and memory parts of the system what to do next. So by strapping together office machinery, Aiken realised that a strange version of Babbage's Engine could be built. Notice that Aiken had extended Zuse's insight – with memory, calculation and control functions all using the same medium.

Aiken proposed the Automatic Sequence Controlled Calculator in 1937. The name tells us a lot about what Aiken saw as special in the machine: it was to be automatic – the human role was minimised to setting up the machine and waiting for the results. In practice, Aiken was to be more pragmatic than his vision of connected punched-card machines suggests, and the designs made use of relays, punched tape, electrical and mechanical parts. It was a neat idea, but would be enormously expensive to build. Big technological projects need powerful allies, and Aiken found two. The United States Navy, aware that rearmament would generate the extreme mathematical challenges Zuse had experienced at Henschel, was persuaded to invest substantially in Aiken's Calculator. With perhaps half a million dollars of Navy money guaranteeing the project – a king's ransom compared to the sum Zuse had begged – Aiken had only to convince IBM to contribute its machines and expertise. The office machine company was won over, not because it wanted to build a scientific computer – for which there was no market – but because links with Harvard created good publicity, and the complexity of the project would stretch the ingenuity of the company's engineers.

Work began at the IBM laboratories at Endicott, New York, and the Automatic Sequence Controlled Calculator was completed two years after Pearl Harbor had made the Navy's interest in fast calculation particularly pertinent. IBM made a gift of the giant five-ton calculator to the university in 1944 in a blaze of publicity, although the director of IBM was furious that Aiken did not give due credit to the company at the inauguration speech. Thus, the Harvard Mark I, as the machine was renamed, became the first 'Robot Superbrain' to be lodged in the public mind. After the ceremony, a veil was drawn over the monster, and urgent, secret calculation commenced: churning out numbers for naval weapon design.

Just as the Harvard Mark I was being unveiled, a second monster was under construction 300 miles further south on the American East Coast. The Electronic Numerical Integrator and Computer – the ENIAC – was also born of speed and conflict. The United States entered the Second World War in December 1941, and committed its massive industrial base. Thousands of state-of-the-art factories, untouched by bombs, would produce the millions of guns, ships and munitions essential to the Allied cause. This industrial might tapped a rich vein of engineering expertise. Not only would more armaments be produced, but new and more deadly designs would multiply their destructive power. However, the deployment of new artillery was threatened by a familiar crisis in calculation.

A 1940s anti-aircraft gun, such as might be installed to protect an aircraft carrier or the city of London, was a complex machine, built of perhaps a ton of steel. Consider how one of these guns had to operate in the stress of sudden attack. A fast-moving aircraft has been spotted, giving the crew, if they are fortunate, a few minutes of preparation. A shell has to be selected and primed. Information about the approaching aircraft – usually a location and an estimate of speed – has to be turned into a prediction of a position along the flight path. Before the gun is directed, other factors have also to be considered: wind speed, the type of shell and the timing of the fuse. Out of all this data, the crew have only a few minutes to make a calculation and get it right. These sums are in fact just as difficult as the simultaneous equations that Zuse was struggling with in Germany. What to do? Brilliant mathematicians were too scarce to deploy in an anti-aircraft crew. The answer was simple but effective: do all the possible calculations beforehand. The anti-aircraft crews were given firing tables: grids of numbers, designed so that as soon as the location and speed of the incoming enemy plane were known, and with knowledge of the other factors, it was a simple matter of looking up in the table where to point the gun. But technology did not stand still. The engineers were churning out designs for ever more powerful artillery, and the improved machines were rolling out of the American factories. Each new anti-aircraft gun required the calculation of a new set of ballistic tables, and the United States Army was increasingly concerned that the moment would arrive when – for lack of mathematics – its new weapons would be useless.

So just as the Navy was persuaded by Aiken of the merits of building monster calculators, so the Army was likewise receptive to the University of Pennsylvania physicist John W. Mauchly when he presented his grand idea in 1941. Mauchly had designed mechanical calculators in the 1930s, and had begun to tinker with electronic versions, which, like Helmut Schreyer's in Nazi Germany, meant valves, or as he would have called them, vacuum tubes.

In 1940, Mauchly had bumped into John V. Atanasoff, a fellow physicist at the far less prestigious Iowa State College. Atanasoff was like an American Zuse: he, too, had been frustrated by simultaneous equations, and had decided, with Clifford Berry, to automate their solution. His 'ABC' – Atanasoff-Berry-Computer – was a little machine, half mechanical, half electronic, resembling the barrel of a barrel organ surrounded by wires (the numbers were stored as minute charges on the ends of the barrel pins). Mauchly seems to have taken away from his meeting with Atanasoff the conviction that a fully electronic calculator was possible (exactly what the ABC did inspire in Mauchly was later the subject of fierce post-war patent litigation).

However, conventional wisdom said that large electronic machines would inevitably fail. The problem was that, like lightbulbs, the filaments of valves eventually burnt out. Think about Christmas tree lights: lightbulbs connected to each other in a long string. The more lightbulbs in the string, the greater the chance that one of them will blow, and the more likely your Christmas will become a time-consuming chore of trying to locate the dead bulb. Mauchly knew that the electronic calculator he was proposing to the United States Army would need thousands of vacuum tubes. The chance of all of them working when the machine was turned on was slim indeed. But the Army was desperate for a fast calculator to produce the firing tables needed at the front, and approved the ENIAC project anyway. Mauchly and his co-designer, J. Presper Eckert, set to work.

The ENIAC was an even larger machine than the Harvard Mark I. It filled a room at the Moore

School of Electrical Engineering, needed 18,000 valves, required a team of operators to tend it, and the anecdote – that when the calculator was finally turned on in 1945, the lights of Philadelphia dipped – may well be true. The Christmas tree light problem had been overcome, partly because Eckert and Mauchly had designed the ENIAC so that faults could be quickly diagnosed and whole blocks of components easily replaced. Partly, conventional wisdom had been wrong: vacuum tubes, if treated carefully, were more stable than originally thought. Ironically, the ENIAC was too late to produce the firing tables it had been commissioned for, but it was soon deployed on other military matters, including early thermonuclear bomb calculations. Like the Harvard Mark I, its success was publicised: a display panel with a bank of lights behind half-ping-pong balls was hastily constructed to make the massive calculator a visually striking focus for cinema newsreel and television. The ENIAC is sometimes called the first computer, but, remarkable though it was, the description is misleading. It was a startlingly fast electronic calculator – it could predict the trajectory of an artillery shell quicker than the shell itself moved. It could also be 'reprogrammed', switching from calculating trajectories to the motion of a shockwave. But this reprogramming meant extensive rewiring: two days spent completely rearranging a nest of plugboard wires. It was not the same machine doing different calculations, but a subtly different machine for each job. The ENIAC was a juggernaut-sized electronic *calculator*, not a universal machine. But the frustration of rebuilding the machine each time the mathematical problem was changed forced the ENIAC team to devise the crucial idea of this book: the concept of the stored program.

We have already seen how both Charles Babbage and Konrad Zuse had hit upon the idea of storing the instructions for calculations in mechanical form, thereby automating the process, and making it simple – and speedy – to switch between mathematical jobs. By 1945, Zuse's projects were, like Germany, in ruins. Mauchly and Eckert and their other collaborators on the ENIAC, located in the intact and prospering America, were in an incomparably more enviable position, and poised to exploit the stored-program idea.

But the impact of war on a third country, Great Britain, must first be understood; since not only did the electronics developed there thwart the ENIAC team's high hopes of being first in the race to build a stored-program electronic computer, but, before Zuse or Mauchly, a third person, Alan Turing, had articulated the stored-program idea, conceptualised in a shatteringly original and profound form.

12   Turing in Secret Britain

Turing's parents hailed from the white administrative elite of the British Raj. His father, Julius Mathison Turing, worked for the Indian Civil Service. The Service had been reformed over the previous century. It often served as a laboratory for new bureaucratic ideas, tried out far from the mother country and enforced by a formidable military presence. Alan Turing was conceived in the port of Chatrapur, near Madras. On discovering the pregnancy, both parents travelled back to England, and Ethel Sara Turing gave birth to Alan on 23 June 1912 in a Paddington nursing home. Julius returned to India, and Mrs Turing stayed to care for her second son, at least until September the following year, when she returned to Julius and India.

This would be the pattern of Alan Turing's early life. While Mr and Mrs Turing were in India, their sons would be farmed out to guardians. Alan spent many of his early years in St Leonards in Sussex, where he was the neighbour of H. Rider Haggard, the author of imperial potboilers *She* and *King Solomon's Mines*. Turing's parents would make the long sea journey rarely, enabling Julius to work his way slowly up the Indian Civil Service hierarchy, and to see his sons maybe once a year. By the time Alan was at the age to attend school, Julius was Secretary to the Madras Government Development Department. Like many offspring from white colonial families, Alan was sent to private boarding schools, which would act both as educator and *in loco parentis*. The first school was Hazelhurst, and during his attendance there he was first exposed to science, partially through popular children's books such as Edwin Tenney Brewster's *Natural Wonders Every Child Should Know*. As Turing's biographer Andrew Hodges has pointed out, books such as Brewster's, although they told simple stories that bolstered the domestic and social life of Belle Époque America and Edwardian Britain, also contained speculations on the nature of mind and machines.

Turing's interest in science increased, although

he had to do well at the more mainstream subjects to pass the entry exam into his next school, Sherborne in Dorset. His day of arrival at Sherborne coincided with the General Strike, an ill-planned and bitter nationwide dispute in May 1926, which brought the trains, along with much of British industry, to a halt. The government beat the strike through violence and volunteer replacement labour. Alan Turing, at the age of 13, beat the strike by cycling 60 miles from Southampton to Sherborne, resulting in an early brush with fame – the feat was reported in the local newspaper, and meant that he arrived at the school with a reputation for eccentricity.

Sherborne School, however, shared with other British private schools a distaste for science, and favoured instead the encouragement of classical languages for the brain, and physical sports, especially rugby and cricket, for the Imperial virtues of manliness, hierarchy and leadership. One of Alan Turing's school reports, for example, concluded that 'if he is to stay at Public School, he must aim at becoming educated. If he is to be solely a Scientific Specialist, he is wasting his time at Public School'. (In a simple, but highly effective piece of educational propaganda, private schools in Britain labelled themselves 'public schools' – depriving the state-funded alternative of the term enjoyed in America.) Mathematics was a part exception to this rule, since it was seen as cultivating the 'tough' analytical mind. Alan Turing was therefore deflected away from the experimental sciences towards the subject that would give him a career. His often self-taught mathematical sophistication increased: he was able, for example, to write a summary and critique of Einstein's special theory of relativity that would have been precocious at university, let alone school, level.

At Sherborne, Alan also developed emotional attachments to other boys, deeper even than was typical at British private schools where homosexuality was practically an institution while remaining illegal in the country at the time. His greatest love seems to have been an older pupil named Christopher Morcom, who shared Alan's interest in science and mathematics. In 1929, both Alan Turing and Christopher Morcom sat the examinations for entry to Trinity College, Cambridge, which had a venerable, if uneven, tradition in mathematics from Isaac Newton in the late seventeenth century to G.H. Hardy, author of *A Mathematician's Apology*, in the twentieth. Christopher was 18 and successful in the exam; Alan was a year younger and not yet ready. He thought he would have a second opportunity to join Christopher at the most prestigious mathematical centre in Britain the following year. However, tragedy struck: in February 1930, before leaving Sherborne School, Christopher Morcom fell grievously ill, suffering complications from bovine tuberculosis contracted as a child from infected milk. Alan Turing knew nothing about the illness until Christopher died within days of taking to bed.

The death of his friend was devastating. But, according to his biographer, Turing's despair was to have profound intellectual consequences. Hodges argues that Turing dwelt on what relationship Christopher's 'spirit' had to the '"mechanism" of the body'. In 1930, he held on to the conventional Christian – or Cartesian – division: the existence of both spirit and the material world. As time passed, Turing was to become more and more convinced that mechanism was all there was, that the mind could be explained merely by appeal to the arrangement of matter. Christopher's death, therefore, confronted Turing with the deep questions of the nature of mind, matter and the difference between life and death.

Alan Turing sat the Cambridge entrance examination for the second time the following year, this time with success. In 1931, he went up to King's College with a scholarship of £80 per year, enough for a spartan undergraduate lifestyle.

King's College is the visual heart of the city: the barn-like Chapel, complete with flying buttresses, overlooks the meadows of the Cam. The view from the other side of the river, with punts in summer, and cows grazing in front of the manicured lawns of King's, is the icon of the University. A tourist walking up from the river through the college to King's Parade passes academic buildings much like those found in nearby Queens', St Catharine's and Corpus Christi. However, the culture within each college varied immensely, and Alan's advent in King's, rather than Trinity, was fortuitous. Whereas Trinity's atmosphere stemmed from its being, by far, the largest and richest of the colleges, King's prided itself on a scholarly celebration of moral and political dissent. The bursar, John Maynard Keynes, contributed the theory of demand management, a radical revolution in economic thinking that only after the Second World War became orthodoxy. To the left of Keynes were communists of various hues. Keynes' personal life was an indicator of King's culture: aesthetically refined, moving in the same circles as the Bloomsbury artists and writers, sharing both the intellectual self-confidence and snobbery of that group, as well as the sexual experimentation. Homosexuality, in private (remember it was illegal) among the elite King's cliques, was a key part of the college culture. If he could gain access, King's, for Alan Turing, had the potential of becoming a rare oasis, sympathetic both socially and intellectually.

The Mathematics Tripos at Cambridge University is taught through often poorly presented lectures, but the college system means that there are many senior mathematicians employed in proportion to the number of students; so an undergraduate, such as Turing, could benefit from occasional first-rate personal supervision. However, Turing seems to have worked largely on his own: in 1934, he thought he had discovered an important new proof in probability theory, only to be told that, while it was certainly profound, the Central Limit Theorem had already been proven 12 years previously.

Figure 5: Alan Turing.

13 Foundations Shaken

Cambridge mathematical education encouraged the best students to pursue their own research, and many were attracted to the intellectual hot spots of the period. The first half of the twentieth century saw a sustained attack on the foundations of the sciences by scientists themselves. In physics, the absolute space and time described mathematically in the late seventeenth century by Trinity College's most famous scholar, Isaac Newton, had been shaken by Albert Einstein's double whammy: the special and general theories of relativity. The special theory of relativity showed that inertial mass and supposedly fixed lengths would vary according to how one moved relative to what was being measured. Einstein was certainly not the only scientist who could have proposed the special theory: hints of a revolution can be found as far back as the 1870s, when Scottish physicist James

Clerk Maxwell proposed his mathematical account of electromagnetic waves. Likewise, the remarkable and beautiful general theory of relativity drew on existing work, in this case that of continental mathematicians of the previous century such as Georg Friedrich Bernhard Riemann (of the Riemann Zeta Function), who had imagined new geometries, ones in which parallel lines might eventually meet. But Einstein applied these new geometries to the natural world, enabling him to bring gravitation in to his scheme, showing that gravitational mass related to the 'curvature' of the combined space-time.

Newton had envisaged the physical world consisting of particles in motion, and he set down the laws governing their movement. Eighteenth-century French natural philosophers had taken the Newtonian mechanics and arrived at an atheistic conclusion that would have horrified Newton himself: give me, argued the Marquis Pierre Simon de Laplace, the positions and velocities of all the particles in the universe, and I will predict the future. The past and future of the clockwork universe were, in principle, knowable and inevitable. In the twentieth century, this secure mechanical vision had also been shaken. Out of the ferment and uncertainty of early twentieth-century Europe came an alternative: quantum mechanics, in which precise knowledge of the position and velocity of any particle, let alone all matter in the universe, was impossible. Like the throwing of pairs of dice, individual events were unpredictable; only when taken together in large numbers would structure and predictability emerge. Newtonian physics had been the steady framework for the rest of the physical sciences, and now that framework had melted away and been replaced by an incompatible mix of relativity and quantum theory.

Nor was the attack on foundations confined to physics. Perhaps the greatest shock occurred in mathematics, the bricks from which the physical foundations of the sciences were constructed. The shift from Newtonian to Einsteinian physics merely demonstrated that different buildings could be made from the same bricks. But what if the bricks themselves were suspect?

14   Mathematics: Truth or game?

An early hint of the troubles to come lay in the new geometry that Einstein had used in his general relativity. For hundreds of years, the model of mathematics was taken from the books of geometry known as Euclid's *Elements*. Euclid was a Greek, active around 300 BC. He gathered together the geometrical knowledge of his time and arranged it in a logical order. Euclid started with a few definitions: a point was something with no length nor breadth; a line was something with length but no breadth; and so on. No problem there. Then he offered a few simple statements that he thought were undoubtedly true, what were called 'axioms'. Most of these were intuitively obvious: 'A straight line may be drawn between any two points'; 'A circle may be drawn with any given point as centre and any given radius'; and 'All right angles are equal'. All seemed sensible. With axioms such as these,

Euclid began to construct proofs, showing that, for example, two of the angles of an isosceles triangle (one with two equally long sides) were equal. What was powerful about Euclid's format was that he started with a small number of simple, true statements, and deduced further theorems until he had covered all of the geometry he knew. As long as the logical steps were persuasive, and you started with truth, then you must end up with truth. True knowledge could in this way be extended reliably and without limit.

In 15 books, Euclid led the reader, truth by truth, to some really complex conclusions: there were only five 'regular solids', objects that had sides made entirely of the same simple flat shapes: the pyramid, the cube, the octahedron, the dodecahedron and the icosahedron (see *Figure 6*). It was not immediately obvious that there should be only five – why not six or 20 or two? But because we could follow Euclid step by step, we knew he must be right. However, Euclid found that for some of the proofs he needed an axiom which was a bit more complicated than the rest: for every line and a point that is not on that line, there exists a unique line that can be drawn through the point such that the

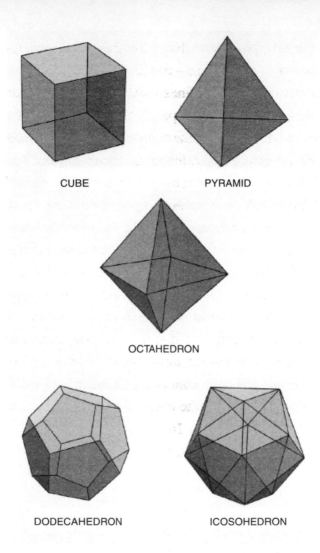

CUBE

PYRAMID

OCTAHEDRON

DODECAHEDRON

ICOSOHEDRON

Figure 6: Euclid's five regular solids of four, six, eight, 12 and 20 sides.

two lines never meet. Now this 'parallel postulate' was not at all obvious – it didn't even look simple, and it talked about lines meeting at a place that might be way off the page. Yet it could not be proved directly from the simple axioms, and Euclid needed it to complete his geometry; so it seemed as if there were no choice but to accept it.

Euclid's *Elements* was studied continuously from the first millennium BC, past the fall of the Roman Empire, through medieval Europe (when very little mathematics was done, outside Islamic lands), through to recent times. It was the staple text of school mathematics, a system of deduction, a model for the rest of mathematics and even the sciences, a beautiful myth. And what is more, Euclidean geometry connected mathematical truths to the physical and the everyday, where they were useful and lucrative. In the nineteenth century, Herman von Helmholtz commented:

> *Conclusion is deduced from conclusion, and yet no one of common sense doubts that these geometrical principles must find practical application in the real world about us.*

He continued:

Land surveying as well as architecture, the construction of machinery no less than mathematical physics, are continually calculating relations of space of the most varied kind by geometrical principles; they expect that the success of their constructions and experiments shall agree with these calculations.

A lot rested on Euclid.

Several attempts were made to tidy up the *Elements* by proving the parallel postulate, or replacing it with something simpler, but with little joy and much frustration. When a young Transylvanian mathematician informed his father that he was considering revisiting the problem, the father urged him to

give it up. Fear it no less than the sensual passions because it, too, may take up all your time and deprive you of your health, peace of mind and happiness in life.

But János Bolyai ignored the paternal wisdom, and, nearly simultaneously alongside other mid-nineteenth-century mathematicians such as the

Russian Nikolai Ivanovich Lobatchevsky and the German Bernhard Riemann, suggested a shocking new approach: replace the parallel postulate with something counter-intuitive – that parallel lines might meet, for example, *but to worry only if a contradiction were found*. If there were no contradictions, these non-Euclidean geometries were just as 'consistent' as Euclid's. This was revolutionary. It wasn't just the invention of exotic mathematics that would interest no one but experts. It suggested something much more frightening: that the Euclidean geometry which seemed to speak truthfully about the physical world was merely one of several possible geometries – so which one, if any, was 'true'? Could you just pick any set of rules, and play with them? Was mathematics a mere game, or was it about truth?

15   Crisis Looms

Non-Euclidean geometries were the first suggestion that mathematics was not merely the steady accumulation of true theorems, and their discovery directed attention towards foundations. Similar anxieties surrounded analysis – the mathematics that underpinned calculus, and therefore also of immense practical significance. As the nineteenth century progressed, the temperature of debate rose. A rearguard action was fought. Some philosophers – such as the Englishmen Bertrand Russell and Alfred North Whitehead, and the German Gottlob Frege – proposed that secure foundations could be found in the surely indubitable rules of reliable thought, and began a mammoth effort to reduce mathematics, including geometry and analysis, to logic. Yet again, paradoxes and problems arose: even a concept as apparently simple as a 'set' turned out to be far more complex than initially hoped. A fin-de-siècle crisis loomed, yet some remained optimistic. In 1900, the Göttingen mathematician David Hilbert addressed the International Congress of Mathematicians in Paris. The turn of the century provoked him to seek hope for the future by looking to the past:

> *History teaches the continuity of the development of science. We know that every age has its own problems, which the following age either solves or casts aside as profitless and replaces by new ones. If we would obtain an idea of the probable development of mathematical knowledge in the immediate future, we must let the unsettled questions pass before our minds and look over the problems which the science of today sets and whose solution we expect from the future. To such a review of problems, the present day, lying at the meeting of the centuries, seems to me well adapted. For the close of a great epoch not only invites us to look back into the past but also directs our thoughts to the unknown future.*

Hilbert then put to his audience of the world's

most distinguished mathematicians 23 unsolved problems. They were not chosen at random but pinpointed the weakest spots in the crumbling foundations. If solved, mathematics would again be secure, and Hilbert was certain they would be:

> [The] *'conviction of the solvability of every mathematical problem is a powerful incentive to the worker. We hear within us the perpetual call: There is the problem. Seek its solution. You can find it by pure reason, for in mathematics there is no **ignorabimus*** [we will not know].

One particular concern of Hilbert's was to be of great significance to Alan Turing and the history of computing. Hilbert described his own philosophy of mathematics as 'formalist', accepting that, as with non-Euclidean geometries, what mattered was that you chose the rules of the game, and as long as the game of mathematics had certain properties, then the mathematics was sound. In particular, Hilbert was asking questions of the axioms of mathematics, ones he stated even more precisely in 1928: given a set of axioms, could we be certain that the system of theorems derived possessed three properties: completeness, consistency and decidability? Mathematics would be *complete* if, starting from the axioms, either a proof or a disproof could be found for every meaningful mathematical statement. There would be no third way, no statements that stood outside what could be proved or disproved. The axioms would be *consistent* if there were no contradictions. If it could be shown by one string of deductions that an isosceles triangle had two equal angles, and by another string that no angles were equal, then there would be something seriously wrong. Finally, was mathematics *decidable*? Given a meaningful mathematical statement, did there exist a method that could be described and applied and would reveal whether the statement was true or false? Hilbert's belief, shared by his audience, was that mathematics must surely be complete, consistent and decidable.

In the year Hilbert retired, 1930, Kurt Gödel, a Czech mathematician in his mid-twenties, began to publish some arguments that shook Hilbert's proposals. Indeed, so shocking were Gödel's conclusions that they were like earthquakes compared to the nineteenth-century tremors of Bolyai and Lobatchevsky. Gödel began with a reasonable set of arithmetical axioms, and began using them to make statements about themselves (crudely speaking, since statements could be numbered, then self-referential statements could be constructed, since the statements themselves were *about* numbers). Then he showed that some of these self-referential statements were analogous to that old logical riddle: 'A Cretan says "All Cretans are liars, and I am lying right now." Is he telling the truth?' That is a perfectly formed grammatical statement, yet we cannot say whether it is true or false. Likewise, Gödel's mathematical statements were meaningful, but could not be proved or disproved from the axioms. Hence Gödel had shown that arithmetic – and therefore mathematics based on a similar set of axioms – must be incomplete (or inconsistent). He had shown that mathematical statements existed that lay beyond proof or disproof.

16   Turing and the Decision Problem

Alan Turing was sitting his university entry examinations when Gödel began publishing his astounding results. The shock was still reverberating when Turing passed his undergraduate degree and was elected a fellow of King's College, Cambridge, at the age of 22. Gödel had, of course, answered Hilbert's questions regarding the consistency and completeness of mathematics, but there remained the issue of decidability. Although the edifice of mathematics was severely tottering, perhaps it could finally be shored up if a definite method could be provided that would decide between true theorems and false statements – perhaps that way the bizarre self-referential assertions of Gödel could be cut out as aberrations, since the method applied to them would reveal whether they were in fact true or not, even if their truth could not be proved with the axioms provided. A throwaway remark by Turing's senior colleague at Cambridge, Max Newman, intrigued him: was there a 'mechanical process' that could be applied?

Newman was using the word 'mechanical' to refer to an unthinking process, describable by an explicit set of rules: do this, do that, and so on. But, 'mechanical' also has a more everyday sense: 'by machine'. This slippage of meaning had profound consequences. Newman's chance remark seems to have percolated for a while within Turing, before the young mathematician hit on the insight: what if 'mechanical' was taken seriously? Or, at least, what if imaginary machines could be applied to the decidability problem? In a whirlwind of activity, begun in Cambridge and completed on a visit to Princeton, Turing worked up his ideas, and presented them in a paper submitted to the London Mathematical Society on 28 May 1936. It was 'read' (accepted) by the Society in November, and published the following year. 'On computable numbers with an application to the *Entscheidungsproblem*' was a remarkable contribution, on three levels. The first level was of the most immediate relevance: he had solved Hilbert's *Entscheidungsproblem*, the issue of decidability.

Turing began by defining his own term: 'computable' numbers. By the twentieth century, mathematicians spoke of many different kinds of numbers. There were the positive integers (1, 2, 3, 4, 5, . . .), the negative integers (–1, –2, –3, –4, –5, . . .), the 'rational' numbers (numbers expressible by a fraction, such as 2/3 or 122/245 or –24/29), and 0. If you put the integers, the rationals and 0 along a line, there would still be some gaps. A number such as the square root of 2 was not an integer, but nor was it a rational (the Greeks had known a proof of this); it was called an 'irrational' number. To fill the line up, you needed all the rational and all the irrational numbers; in total these made up the 'real' numbers. Turing defined his computable numbers as 'the real numbers whose expressions as a decimal are calculable by finite means', or, in other words, whether there was a mechanical method that could be applied that would generate the decimal digits of that number. For example, '1/3' was a computable number because by using the long division taught in schools – a mechanical method, just turn the crank – you generate first 0.3 then 0.33 then 0.333, and so on. The square root of 2 was a computable number, because you can follow an explicit method that would spit out its infinite decimal representation (first 1 then 1.4 then 1.41 then 1.414 then 1.4142, and so on).

Turing then imagined a machine that could be used to generate computable numbers. His description of them is complex and mathematical, but the following gives a gist of his style:

We may compare a man in the process of computing a real number to a machine which is only capable of a finite number of conditions q_1, q_2, ... q_R which will be called 'm-configurations'. The machine is supplied with a 'tape' (the analogue of paper) running through it, and divided into sections (called 'squares') each capable of bearing a 'symbol'. At any moment there is just one square, say the r-th, bearing the symbol $T(r)$ which is 'in the machine'. We may call this square the 'scanned symbol'. The 'scanned symbol' is the only one of which the machine is, so to speak, 'directly aware'. However, by altering its m-configuration the machine can effectively remember some of the symbols which it has 'seen' (scanned) previously. The possible behaviour of the machine at any moment is determined by the *m*-configuration q_n and the scanned symbol $T(r)$. This pair q_n, $T(r)$ will be called the 'configuration': thus the configuration determines the possible behaviour of the machine. In some of the configurations in which the scanned square is blank (i.e., bears no symbol) the machine writes down a new symbol on the scanned square: in other configurations it erases the scanned symbol. The machine may also change the square which is being scanned, but only by shifting it one place to the right or left. In addition to any of these operations the *m*-configuration may be changed. Some of the symbols written down will form the sequence of figures which is the decimal of the real number which is being computed. The others are just rough notes to 'assist the memory'. It will only be these rough notes which will be liable to erasure. It is my contention that these operations include all those which are used in the computation of a number.

In other words, Turing pictured a *symbol-manipulating machine* which could be said to have a certain number of states (the '*m*-configurations'), and it repeatedly scanned a paper tape, and then either printed something on that tape, or erased the symbol already there, and could then move left or right, and then change its state – in all cases in a manner *determined* just by two things: what state the machine started in and what symbol it saw.

He gave some examples of these simple 'computing machines', beginning with one that could spit out the sequence 0 1 0 1 0 1 0 1 0 1 ... Turing noted that the entire action of a computing machine could be expressed as a table, which listed for every state what the machine would do if it saw each symbol. Each number had at least one machine, and each machine had a table.

Turing then discussed ways that all the different sorts of table imaginable could be put into a standard form, so they could be compared. From there it was a short step to assign every machine a number. He showed a method, for example, for attaching to the machine that generated the sequence 0 1 0 1 0 1 0 1 0 1 . . . a 'description number', in this case '3133253117311335311173111332253111173111133531731323253253117'. That's a very long number, but it was uniquely attached to a certain machine. The great fact about numbers is that they can be put into an order. Therefore Turing had a way of ordering all the machines that produced computable numbers.

By imagining these machines, Turing had taken an important step: he had given a straightforward and easily comprehensible meaning to the mathematical concept of 'decidability'. Now any Turing machine was either 'good' or 'bad'. A good Turing machine would spit out a sequence of numbers. However, in the course of their operation, some machines fell into an infinite loop – they got stuck. (Think about how your computer goes wrong. Unlike a car, for example, it is apt to 'freeze' when it goes wrong. The computer is still 'working', but it is trapped, repetitively cycling through a small loop of instructions.) Turing's crucial question was to ask whether it was possible to build a machine, let's call it 'D', that would decide, given the description number of any particular Turing machine, whether it was good or bad. It might, for example, print 'Yes' if the Turing machine was good, and 'No' if the machine was bad. Let us assume, Turing said, that D does exist. Turing then imagined what he called the 'universal computing machine', essentially a single Turing machine that could imitate all other

Turing machines, to take their description numbers, decode them to work out what they did, and then duplicate their actions. 'It is possible', he wrote, 'to invent a single machine which can be used to compute any computable sequence. If this machine U is supplied with a tape on the beginning of which is written the S.D. [Standard Description] of some computing machine M, then U will compute the same sequence as M.' He outlined the behaviour of the universal machine, and even gave the table which described it.

Now imagine we chain up the deciding machine D with the universal machine U, and call this combined machine 'M'. Since M is a Turing machine, it must have a description number. What happens, Turing asked, if we feed this number back into M? First it goes into the deciding machine D, and it must spit out 'Yes' – it can't be a bad machine because we are assuming D exists and we can write a table out for U, so that definitely exists too. But when the second part of the combined machine, the universal part, takes the description number, *it must start simulating itself*. Indeed, Turing argued, it must get into an infinite loop. It must be trapped. The answer spat out by D should have been

'No'. We have, said Turing with a flourish, a contradiction. He had found an example of an undecidable mathematical statement, and a counterexample to Hilbert.

Turing's style of argument had a pedigree. Like Gödel's – and like Georg Cantor's diagonal argument (see pages 94–6) – it worked because of self-referentiality. Gödel had numbered statements about numbers. Turing had machines simulating themselves. Both of these twentieth-century logicians were indebted to their nineteenth-century predecessor. Georg Cantor was a German mathematician born in St Petersburg in 1845. At Berlin he had been taught by a great trio: Ernst Eduard Kummer, Leopold Kronecker and Karl Weierstrass, who had been driven by a desire to make analysis rigorous. The problems with analysis stemmed from the use of the infinite: just what did an infinitely small quantity mean? Intuitively it is easy to guess what the infinitely small must be like: just start with something small, and divide it again and again and again. But intuition lays traps for mathematics, just as it is essential for breakthroughs in the first place – indeed, much of mathematics can be understood as an uneasy dance, alternatively

Georg Cantor's Diagonal Argument (1873–74)

(This version is adapted from Hodges, 1983. Cantor's 1874 paper presents the argument in more abstract terms.)

Write down a list of rational numbers between 0 and 1.

For example, we could cover all of them by starting as follows (take the halves, then the thirds, then the quarters, and so on, but omit duplications – for example, two quarters):

1/2, 1/3, 2/3, 1/4, 3/4, 1/5, 2/5, 3/5, 4/5, 1/6, etc.

Now write them out in decimal form:

1	0.500000000 . . .
2	0.333333333 . . .
3	0.666666666 . . .
4	0.250000000 . . .
5	0.750000000 . . .
6	0.200000000 . . .
7	0.400000000 . . .
8	0.600000000 . . .
9	0.800000000 . . .

10   0.166666666 . . .
etc. etc.

Notice that every rational number between 0 and 1 will have a numbered position on the list. This means there are as many rationals as there are positive integers. Cantor said that this meant the infinity of rationals was the same infinity as that of positive integers. Mathematicians called this infinity 'Aleph Nought', $\aleph_0$.

Now consider the number formed by taking the first decimal digit of the first number, the second decimal digit of the second number, and so on: 0.5360000006 . . .

Form a new number by adding one to every digit (with the additional rule that if we add one to nine, we get zero): 0.6471111117 . . .

Is this number on the list?

No. Because it differs from the first number by the first decimal digit, it differs from the second number by the second decimal digit, and it will differ from the 9456th number on the list by the 9456th digit. We have made a number that is not on the list. All the rationals are listed, so the number must be irrational. Since it is always possible to do this, no matter how you construct the list, then (unlike rationals and the positive integers) the irrationals – and therefore the reals – are unlistable. The infinity of irrational numbers between 0 and 1 is greater than $\aleph_0$, and was called 'c'.

Cantor thought he had shown that it was possible to talk coherently about different sizes of infinity. Other mathematicians, notably the powerful Kronecker, vehemently disagreed. Attacks by Kronecker contributed to Cantor's mental collapse in 1884. Cantor also thought that the 'transfinite' numbers – those exceeding all finite numbers – had been revealed to him directly by God (he was brought up a strict Lutheran), and his discussion of the infinite(s) upset theologians too.

(Cantor suspected, but could not prove, that there was no infinity lying between $\aleph_0$ and c, what is known as the continuum hypothesis. Kurt Gödel in 1938 showed that the continuum hypothesis was an example of mathematical incompleteness: it could not be disproved using the axioms of set theory.)

embracing and rejecting intuition. Weierstrass set out to provide an account of analysis without any reference to infinitely small quantities. Cantor began by following Weierstrass's interests, but went about the journey in a radically different way. For Cantor, the fundamental object was the 'set', an arbitrary collection of objects, a very general concept. All the integers – as well as the members of a football team or all the three-legged dogs – are examples of sets. Some sets are finite (the football team has 11 members) and some are infinite (the integers). In the 1880s, Cantor presented an argument concerning infinite sets that was simple but stunning: there were different 'sizes' of infinity. In particular, the number of irrational numbers was a 'larger' infinity than the number of rationals. Cantor's argument – like Turing's and Gödel's (and indeed Zuse's calculating machines) – depended on something important: a blurring of the distinction between data and instructions. The number 7, for example, is first taken to be part of the instructions (find the 7th number on the list), but then the argument turns on '7' also being part of the data ('7' is a positive integer). What Turing recognised in Cantor's method was this equivalence of media: both instructions and data having the same form.

So what had Turing found? His machines imitated the mechanical processes of mathematics. He knew from Gödel that there existed mathematical statements that could be neither proved nor disproved. What Turing had shown was that it was impossible to *decide* by applying a mechanical process whether those mathematical statements were true or false. Hilbert's dream of shoring up the foundations of mathematics by appeal to the processes of moving from mathematical statement to mathematical statement was shattered.

Remember I said that Turing's 1937 paper possessed many levels? The first was that he had answered Hilbert's decidability question in the negative, an important result, but of comprehension – and even interest – to only a handful of mathematicians in the world. In fact, an American, Alonzo Church, had just pipped Turing to the post: just as Turing was submitting his proof to the London Mathematical Society, Church published two papers on what he called 'effective calculability' in the *American Journal of Mathematics* and the *Journal of Symbolic Logic*. Effective calculability was equivalent to Turing's notion of computable numbers, and Church had indeed shown that it, too, answered Hilbert's question. Before his paper was finally published, Turing slipped in an appendix noting the similarity. But Church's approach was relatively conventional; it did not proceed via anything like Turing's radical reinterpretation of 'mechanical' methods. In constructing his proof, Turing had imagined, almost as a mere step in the proof, a new sort of machine. Most of his machines were single-purpose: for example, the one that generated a specific number (such as 0 1 0 1 0 1 0 1 . . .), but to make his argument he needed a machine that could imitate all these machines. The imagining of the universal computing machine was the second level of significance of the 1937 paper.

In the next few years, Turing, still a young man, shuttled back and forth between Princeton and Cambridge. At Princeton, he was at the Institute for Advanced Study, a hothouse for the intellect that benefited greatly from the flood of talent escaping Fascism. Einstein was there, as was the prodigy John von Neumann, although Turing missed meeting Kurt Gödel by a year. At Cambridge, he encountered Ludwig Wittgenstein, and sat in on some of the philosopher's seminars. The similarities and, more to the point, contrasts between the two should have made sparks fly. Both were incisive, radical logicians; both made profound appeals to the serious nature of games; both undermined the foundational projects of their disciplines, mathematics and philosophical logic. Their social backgrounds were poles apart, Wittgenstein a son of the aristocracy, Turing a colonial administrator's son. The two potentially had much to talk about, but as surviving transcripts of the ensuing discussion show, they seem to have skirted around each other. Perhaps there was not enough time.

17   Government Codes

In September 1939, the Second World War broke out when the German Army occupied Poland. On the 4th of that month, as international tensions were reaching a peak, Alan Turing arrived at the Government Code and Cypher School at Bletchley Park. It may be that he had been contacted as far back as 1936; certainly in 1938 Turing had visited the British code-breaking establishment and had been introduced to its work. In that year, Bletchley Park was more like a Cambridge college than a military site. A handful of code-breakers, working largely with pen and paper, attempted to keep track of the burgeoning number and types of coded messages emanating from the continent. During the First World War, the British armed services – especially the Admiralty, since ships were beginning to make extensive use of radio – had built up fairly formidable and secret code-breaking facilities. On the staff of Room 40, the naval centre named after its office in Old Admiralty Buildings in Whitehall, was A.G. 'Alastair' Denniston. In 1919, the naval and army signal intelligence units had merged, and, with Denniston as its head, a peace-time code-breaking service had been formed with a public remit to 'advise as to the security of codes and cyphers used in all Government departments and to assist in provision', and with a secret additional term of reference 'to study the methods of cypher communications used by foreign powers'.

The foreign power of most interest to the British was not Germany, whose codes could not be broken, nor the United States nor France (whose codes were cracked by Government Code and Cypher School from 1921 and 1935 respectively), which were targeted because they were economic competitors – but the ideological enemy, Soviet Russia. The code-breakers initially had considerable success cracking the Bolshevik messages, since in cryptography – the science of writing codes – at least, there had been no revolution in 1917, and the old Tsarist methods continued. But in 1927, the Prime Minister Stanley Baldwin stood up in Parliament, and what he read out shocked British cryptanalysts. Baldwin had been infuriated by activities of Soviet diplomats in Britain, who had been travelling around the country promising Kremlin gold to British communists. Baldwin decided that a public shaming of the diplomats might embarrass the Soviet Union into desisting, and to achieve this he read out decryptions of revealing Russian telegrams. The Soviet Union promptly upgraded the security of its codes, leaving the Government Code and Cypher School helpless. This was a definitive moment in the culture of code-breaking secrecy in Britain. Since 1927, the Government Code and Cypher School, and its successor GCHQ, has guarded its knowledge, methods and even existence to an extreme degree. This intense secrecy has had many effects, one being a severe limitation on how reliable our knowledge is of the history of British computing.

What we do know of the British code-breakers is that by the mid-1930s they were complaining of underfunding and poor morale. However, German rearmament led to a revival of interest, more funding, better facilities and a plan for recruitment if tensions turned to war. Recruitment networks centred on the Cambridge colleges, especially

Sidney Sussex and King's, home to Alan Turing. The rearmament also caused a shift of target, away from Bolshevik Russia towards Germany, where the main tool of encryption was the Enigma, a machine that resembled a large typewriter and which converted readable text into a jumbled incomprehensible sequence of letters. Once a message was passed through the machine, it seemed that only someone with another Enigma could possibly read it again. Enigmas were a commercial product; they had been displayed for sale in 1923 at the International Postal Union congress. The trick was that just possessing an Enigma was not enough, however. On top of the 'typewriter' was a series of wheels which could be set in a large number of different positions. Variants of the Enigma were used by the German Navy, Army, Air Force, railways and *Abwehr* (the secret intelligence service of German High Command) to code messages. It worked by passing an electric current through a series of rotating wheels and a wired plugboard, so that each letter was encoded: if 'A' was pressed then 'X' might light up, but if 'A' was pressed again, one of the wheels would turn, and the new encrypted letter might be

'Q'. These wheels therefore added an extra jumbling of the text. In one more act of complication, the wiring within the machine could be changed by using a plugboard, and this gave the encryption process a final twist. The result was that even if you possessed an Enigma, you were helpless unless the position of the dials and the internal wiring were known. German cryptographers considered Enigma invulnerable.

However, a remarkable stroke of good fortune transformed the Government Code and Cypher School's attack on Enigma. In 1931, German traitor and *bon viveur* Hans-Thilo Schmidt passed documents relating to Enigma, including diagrams of its internal mechanisms, to the French secret service. When the French concluded that Enigma remained unbreakable, they passed the documents to the Polish cryptographer Marian Rejewski. By 1933, Rejewski had succeeded in decrypting some Enigma messages, and his team spent the next few years developing their methods, including building an electromechanical device they called a 'bombe' to assist in the deduction of the Enigma wheel settings. Only months before war would be triggered by the German invasion of Poland, in July 1939, the Polish intelligence service met a party of their French and British counterparts in the Pyry forest outside Warsaw and passed on their methods, a reconstructed Enigma and the bombe.

That was the position when Turing arrived at Government Code and Cypher School at Bletchley Park in September. There are many stories about the character of the 'Prof': his chaining of his coffee mug to a radiator, his wearing a gas mask through the summer to ward off the pollen that induced his severe hay fever, and other eccentricities (enough to distinguish him as unusual in a Park that had many eccentrics). On the personal front, there were strange developments, too. The mass mobilisation of the Second World War meant that men and women found themselves working together in unusual situations. There were land girls on the farms, women in the munitions factories, and large numbers of young female volunteers at Bletchley Park. (In a revealing entwining of class and security attitudes, many of the women allocated to British code-breaking work came from well-to-do family backgrounds; it was not until the 1950s, with the exposure of the Cambridge ring of Soviet spies, that this equation of the upper class with trustworthiness in matters of national security would be challenged.) Most of the new staff – and Bletchley Park employed thousands by the end of the war – were clerical. However, women mathematicians also numbered among the civilian elite at the Park, although they were not granted equal pay and status. In the spitfire summer of 1940, Joan Clarke was recruited and assigned to work with the team tackling naval Enigma. The following year, Turing and Joan became close, a relationship that deepened to the point of engagement, although Alan had confided that he knew himself to be homosexual. There was no marriage, although given the prejudices of the age, it would not have been unusual for a gay man to marry for the sake of respectability (and friendship).

Bletchley Park was organised by 'Huts', named after the makeshift assemblage of buildings hastily erected in the grounds of the old mansion to house the clerical armies of the code-breaking project. While the low buildings looked chaotic, they concealed an information-processing organisation that was as sophisticated in its emphasis on speed of throughput as Henry Ford's most efficient factories. Listening stations scattered across

Britain (and abroad) recorded high-speed German transmissions, and relayed the raw data to Bletchley Park. Once there, it flowed in two separate streams. Messages encoded by the Red Enigma, used by the German Air Force, went to Hut 6, and, if decrypted, passed on to Hut 3 where they were then translated and their importance assessed. In parallel, naval Enigma messages passed first to Hut 8, under Alan Turing, and thence onto Hut 4. The cream of Huts 3 and 4, 'Ultra', went straight to Winston Churchill. The authority on the military use of Ultra has concluded that whereas the Ultra intelligence did not win the war (it was not really used until after the United States had entered the conflict, from which point an allied victory was assured), it did decisively disrupt the U-boat campaign, it tipped the balance in the Allied landing on the Normandy beaches, and, in total, it shortened the Second World War by several years.

Much ingenuity went into the attack on the raw undecrypted messages in Huts 6 and 8. A mistake by a German Morse operator – for example, repeating an opening signal or forgetting to change the Enigma dials – was pounced upon and exploited. Other methods searched for tell-tale statistical regularities that might shorten the odds. As the situation of the war turned grave in 1940, and shipping losses in the Atlantic and Mediterranean to the U-boats reached devastating levels, the pressure on Bletchley Park became intense. Code-breaking became less and less a matter of brilliant individual virtuosity, and more and more an industrial enterprise. Staff numbers increased (peaking at 8,995 in 1945), and the work was mechanised. Sometimes, the romantic impression is given that Bletchley Park was a collegiate gathering of geniuses. In fact, it was far more like a munitions factory than a Cambridge college, and even the geniuses worked to make it so. Turing's early job had been to improve the bombes, the machines that increasingly replaced the human processes of guessing decryptions. Great numbers of bombes were built by the office machinery company British Tabulating Machines (owner of Hollerith's licence in the British Empire), and installed in the Bletchley Park production line.

In the United States, the spur of war and the crisis over ballistic tables had prompted the construction of the ENIAC. War demanded speed, which in turn meant electronics. In Britain, the problem was not fast production of guns, but fast production of information. And again, the intense demand for speed led to electronics. As the crucial theatre shifted from the Atlantic to the European continent, the codes used extensively on land became relatively more important. One of these codes, named 'Fish', applied to teleprinter signals. Teleprinters were fast and automatic, and although developed as recently as the 1930s, their use was already expanding as the pressures of war increased the amount of information circulating. Great quantities of 'Fish' messages were collected and relayed to Bletchley Park. Again there was a bottleneck, and attention turned towards innovating methods to attack it. Turing had contributed a statistical technique, which led to a small but steady stream of decrypts from 1942. But faster methods were needed. High-speed machines, able to assist the comparison of messages, were rapidly constructed. So rickety in appearance were these machines that some were nicknamed 'Heath Robinsons'. They did not meet the demand.

In 1943, the mathematician Max Newman, Turing's older colleague at Cambridge who was now in charge of many of the Bletchley Park machines, proposed a radical solution: a machine named 'Colossus' that would store sequences of symbols and compare them at electronic speed. The formidable electrical engineering task of designing and building a machine using 1,500 valves was tackled by a team under T.H. Flowers at the Dollis Hill post office research station in North London. The ten Colossi that were constructed between 1943 and 1945 were remarkable machines: they tried out encryption patterns and compared them to encoded texts until they found a similarity. The machine interacted with the human operator, stopping for advice at certain points. As Jack Good recalled later, there was a 'synergy between man, woman and machine, a synergy that was not typical during the next decade of large-scale computers'. The Colossi even made 'choices' internally, deciding according to pre-set instructions which code-breaking path to follow. This meant that a Colossus had a limited 'programmability', but it was not a computer; it did not have the full flexibility of a universal machine.

Turing was not directly involved with the design of Colossus, although it was automating some of his anti-Fish techniques. But it was a symbol-processing device that immediately recalled the imaginary machines he had conjured up in his 1937 paper. With the bombes, Robinsons and Colossi, he was surrounded by machines that stored and compared and wrote symbols. Indeed, Bletchley Park as a whole was an organised 'machine', taking in raw data and generating 'intelligence' through the repetitive but essentially mechanical processing of symbols. As the Second World War neared its close, conversation with his fellow mathematician Good and Classics scholar Donald Michie turned towards the mechanical nature of intelligence: the questions of what a machine could do. Could it do anything that a brain could do? Could it learn? Could it think?

18 The Computer

The first electronic stored-program computer was
built in Manchester and ran its first program in
June 1948. Yet so far, the industrial city in north-
west England has barely entered the story at all.
The link between the devastated world of 1945 and
the first computer three years later is a fine illustra-
tion of the importance of contingency in history.

By the 1930s, the cotton trade that had made
Manchester prosperous in the nineteenth century
had receded. The city now depended upon ware-
housing and light and heavy engineering, with its
symbol no longer the cotton mill but the Trafford
Park industrial estate, home of Ford and
Metropolitan-Vickers. The University's strengths,
engineering and physics, tied it to the city. (The
latter depended on large electrical devices such
as Van de Graaf generators and early particle

accelerators, so industrial links were crucial.) In 1945, patches of the city were burnt-out buildings and rubble, the consequences of the heavy bombing raids of Christmas 1940. Manchester therefore resembled many European cities. It was not, however, pulverised like Berlin – in that sense it was a more likely location for the first computer than Konrad Zuse's hometown.

But in the United States, factories remained unmolested by bombs and strained at full production. The research and development laboratories of IBM in upstate New York could reflect on their Harvard Mark I, while, also on the East Coast, the United States military and academics at the University of Pennsylvania possessed the ENIAC, the fastest electronic calculator in the world. In the United States were the skills, the knowledge, the money and the institutions with which to build the first computer. So why Manchester? The answer is that the Manchester machine was essentially an American concept, with the addition of one home-grown – and crucial – technique.

The Second World War had mobilised many academic scientists, mathematicians and engineers, and in 1945 they returned to the universities.

That year saw two highly significant additions to Manchester University: a group of mathematicians, including Max Newman and Jack Good, travelled from Bletchley Park; and a group of electrical engineers, including F.C. 'Freddie' Williams and Tom Kilburn, trekked from the radar centre at Malvern, the Telecommunications Research Establishment. Both groups could easily have gone elsewhere. The intense secrecy of British code-breaking meant that the engineers would have known nothing about the Colossus or other Bletchley innovations. Newman, inspired by the Colossus, however, wanted to develop the techniques of electronic logic further, and secured an unusual and sizeable grant from the Royal Society to fund the construction of a machine.

Recall that at this moment the ENIAC team, frustrated that their machine had to be rewired each time the calculation changed, had devised the idea of a stored program: the storage of a list of instructions internally within the calculator, thus adding speed and flexibility. This idea was written up as a proposal for a new machine, the *Draft Report on the EDVAC*, published under John von Neumann's prestigious name, but largely the work of

J. Presper Eckert and John Mauchly. The Moore School of Electrical Engineering, home of the ENIAC, held a summer school in 1946 where the ideas of the EDVAC and the stored-program computer were presented. Newman sent one of his Manchester mathematicians, David Rees, to Philadelphia to attend. Rees passed his notes to Newman and Turing. Both would have immediately recognised the EDVAC stored-program concept as similar to the ideas that had been kicked around Bletchley Park several years earlier – but the unprecedented security regime clamped over the code-breaking innovations meant that no admission of this fact could be made. Other Britons were at the summer school, including Maurice Wilkes, who returned with the stored-program concept to Cambridge, and Douglas Hartree, the authority on numerical methods of solving mathematical problems who had constructed an analogue calculator, a Differential Analyser, in the 1930s. Hartree, an expert on machine calculation, was advising the ENIAC team and ensured that the links with the British were maintained. He had arranged for Freddie Williams, in the United States on a radar assignment, to see the ENIAC in 1946, and had strongly supported Newman's application for the Royal Society grant.

Through these transatlantic contacts, the EDVAC model of the stored-program computer fed back to Britain. At Cambridge, the Bletchley Park model was (ironically) of no influence, and Wilkes set out to build a purely American-inspired machine. This was reflected in Wilkes' name for his stored-program computer: the EDSAC, with its direct echo of Eckert and Mauchly's EDVAC. The least American influence was felt in London, at the National Physical Laboratory, where the recently installed Alan Turing planned a computer along his own lines, the Automatic Computing Engine, or ACE. The mingling of the two influences – although all the time hampered by security restrictions on the mathematicians' side – was at Manchester University. While remaining independent, the mathematicians and the engineers shared a conviction that a stored-program computer could be built.

However, the biggest advantage that the Manchester engineers possessed was technological: techniques for storing and manipulating electronic pulses, skills that came directly from Williams and Kilburn's radar experience. Unlike any of the other teams in the world, the Manchester engineers had the components of a working electronic memory. The 'Williams Tube', as it became known, was built around a type of large mass-produced valve, the cathode ray tube, found in oscilloscopes and televisions (and modern computer monitors). If you turned off an old television set, you would get a bright dot in the centre of the screen that would slowly decay. This slow decay was the key: that point on the screen can be considered as storing a bit of information, albeit briefly. If you can find a way of detecting that dot, and refreshing it, so that it did not decay, then you had the means of making an electronic memory. This is what Williams, with his radar skills, could figure out. He decided to build a small experimental computer around his invention, and the result, after two years of coaxing, was the first operational electronic stored-program computer in June 1948. Williams recalled the moment:

When first built, a program was laboriously inserted and the switch pressed. Immediately, the spots on the display entered a mad dance. In early trials it was a dance of death leading to no useful result, and what was even worse, without yielding any clue as to what was wrong. But one day it stopped and there, shining brightly in the expected place, was the expected answer.

At that moment his universal machine was unique.

19 Minding the Gap:
Many universal machines

At the beginning of this book I said that computers presented a strange case in the history of technology, and asked how their invention might be explained. I added that this was the same question as asking what sort of society would ever need such a thing as a universal machine. The story I have told so far has moved back and forth, reaching as far back as Euclid and as far forward as Manchester in 1948. I have recounted the immediate circumstances that inventors such as Turing, Zuse, Mauchly and Williams worked in, the materials they had to hand, and some of the pressures that pushed them. The story after 1948 is obviously one of unprecedented multiplication of machines, so that the unique machine of 1948 can now be found replicated (in a myriad of forms, mostly much smaller) across the world. Just explaining the one-off invention of a new machine is not enough, since every time it is reproduced calls for a story to explain why it was useful in those differing circumstances. So how did the universal machine come to be used universally, and in what kind of world?

When Turing arrived in Manchester in 1948, he thought he was joining Newman's group. However, the engineering success of Williams' team had transformed the situation. Turing had left the National Physical Laboratory because he was frustrated by the slow pace of development of the ACE, and because his ideas for its design had been sidelined. At Manchester, he was too late to shape the design of the computer, but at least he could work with a machine that was up and running. Turing therefore found himself in a difficult position, on the edge of an engineering culture that had little sympathy for his philosophical investigations, however profound, but also in a position to immediately investigate the practical problems of programming. The Manchester machine was slowly upgraded, and by 1949 it had a decent-sized memory. It was still an experimental machine, though: a mass of wires and valves. At this moment, negotiations began with a local electrical engineering firm, Ferranti, to turn the experimental monster

Figure 7: The experimental Manchester computer, 1949. This picture shows the first electronic computer one year after the first stored-program was run. The crucial Williams Tubes are on the right-hand side. (Source: National Archive for the History of Computing.)

into a sleek commercial product. The University received the first of these in 1951. It ran until the late 1950s, and it was this machine that the technicians labelled the 'Blue Pig'. Turing wrote the programming manual for it (and unlike any other programming manual, it is a good read).

With the hiatus over the change of machines, Turing was able to publish his thoughts on machine intelligence. His biographer, Andrew Hodges, has insightfully traced Turing's development of his

Figure 8: The experimental Manchester computer, 1949. In the centre are the senior members (F.C. Williams and Tom Kilburn), surrounded by their white-coated team (from right to left: Dai Edwards, Alec Robinson, G.E. Thomas). (Source: National Archive for the History of Computing.)

machine-mind ideas from childhood. In published form, the hints are there in his 1937 paper. In it he turned from his imaginary machines to consider the work of a human 'computer' (which, in the 1930s, merely referred to someone who computed, usually repetitive mathematics). His immediate aim was to show that this human computer was doing nothing that his machines could not do, thereby bolstering his argument that through his machines he was addressing all mathematics, that his '"computable" numbers include all numbers which would naturally be regarded as computable'. However, the easy transition from the work of the human computer to the work of the machines suggested an equivalence that, although not explicitly developed, was certainly present: the 'behavior of the [human] computer at any moment', Turing tells us, 'is determined by the symbols which he is observing, and his "state of mind" at that moment'. If so, the entire actions of the human computer, his mind whilst working on the mathematical problem at hand, could be represented by a Turing table (and therefore imitated by the universal machine).

At Bletchley Park, he had been surrounded by machines of many kinds, some making decisions, some storing and manipulating symbols in electronic form. He had discussed machine learning with Michie, Newman and Good. He had made further notes on intelligent machinery at the National Physical Laboratory, where they went unpublished and unappreciated. His free role at Manchester allowed him the time and space to write them up. Alan Turing's paper on 'Computing intelligence and machinery' appeared in the philosophical journal *Mind* in 1950. It is as accessible and readable as the 1937 paper was arcane and difficult. 'I propose', he began, 'to consider the question, "Can machines think?"' His answer was not a direct 'yes' or 'no'. Instead, he showed a way that the question could be answered.

Turing proposed what he called an 'imitation game'. Imagine an interrogator is connected, ideally by a teleprinter, to a man and a woman in another room. Could the interrogator, by asking questions of the man and the woman, guess which was of which sex? (Of course Turing was an expert in decrypting messages over teleprinters, and, as a homosexual in a homophobic world, no stranger to imitation games.) Now what would happen if a machine replaced the man? What would be the conclusion if the interrogator could not tell the difference between human and machine? Turing asked: 'Will the interrogator decide wrongly as often when the game is played like this as he does when the game is played between a man and a woman? These questions replace our original, "Can machines think?"' The set-up makes human and machine communication comparable ('We do not wish to penalise the machine for its inability to shine in beauty competitions, nor to penalise a man for losing in a race against an aeroplane'), yet it still gives the human the advantage – the machine is trying to imitate the human, not the other way around – which makes it a stronger test. The machine Turing had in mind was the digital computer: 'these machines are intended to carry out any operations which could be done by a human computer. The human computer is supposed to be following fixed rules' (note the equivalence implied with the universal machine from 1937). His 'belief' was that the digital computer would give a 'good showing' in the imitation game. Turing even offered a prediction:

I believe that at the end of the century the use of words and general educated opinion will have altered so much that one will be able to speak of machines thinking without expecting to be contradicted.

'Can machines think?' should perhaps have been subtitled 'Nine arguments against doubters', since Turing spends most of the paper dismissing all possible objections. He showed that machines could do the unexpected, or could imitate other machines, and even considered the effects of Extra-Sensory Perception (ESP), should it ever be discovered. Ironically, given the rather scholastic form of the paper, the first objection dealt with was theological: 'Thinking is a function of man's immortal soul. God has given an immortal soul to every man and woman, but not to any other animal or to machines. Hence no animal or machine can think.' He had little patience with such views, complaining sarcastically: was it not a 'serious restriction on the omnipotence of the Almighty' to state that God could not grant machines souls? Just as Galileo's attackers had cited Joshua ('And the sun stood still') to denounce Copernicanism, present-day theologians would be proved ignorant in the long term if they opposed the machine mind. Indeed, Turing viewed the Theological Objection as a mere variation on the 'Heads in the Sand' Objection: 'The consequences of machines think-ing would be too dreadful. Let us hope and believe that they cannot do so.' Such opponents, Turing suggested rather archly, should perhaps seek consolation in the possibility of the transmigration of souls. However, in passing, he made a rather interesting observation: Ostrich-like attitudes were often 'quite strong in intellectual people, since they value the power of thinking more highly than others, and are more inclined to base their belief in the superiority of Man on this power.' One root of the snobbishness shown to the late-twentieth-century computer nerd, perhaps?

The remaining objections were more weighty. Did Gödel's demonstration of the limitations of logical systems, ones shared by any digital computer, mean that there was an essential difference between humans and machines? No, since we do not know that these same kinds of limitations do not apply to humans too. Then there was the 'Argument from Consciousness', although it would be better labelled the 'Feeling Objection'. In 1949 the Lister Oration on the 'Mind of Mechanical Man' by Dr G. Jefferson had been published by the *British Medical Journal*. Turing extracted one passage:

Not until a machine can write a sonnet or compose a concerto because of thoughts and emotions felt, and not by the chance fall of symbols, could we agree that machine equals brain – that is, not only write it but know that it had written it. No mechanism could feel (and not merely artificially signal, an easy contrivance) pleasure at its successes, grief when its valves fuse, be warmed by flattery, be made miserable by its mistakes, be charmed by sex, be angry or depressed when it cannot get what it wants.

Of course, the Blue Pig had expressed its yearning in love poetry, but this was precisely the 'easy contrivance' that Jefferson ruled out. But as Turing pointed out, this was equivalent to arguing that you have to *be* someone else to *really* know that they feel. The world could be all a simulation, everyone else could just be pretending to feel – a solipsistic argument that applied to other humans as well as machines. And who knows, perhaps the Blue Pig did feel grief when its valves fused.

If sceptics were not convinced when a machine wrote poetry, they tended to search around for another disability: it could never be kind, or have a sense of humour, or make mistakes, or fall in love, or enjoy strawberries and cream. As Turing noted,

Figure 9: The Ferranti Mark I, also known as the Blue Pig, 1951. (Source: National Archive for the History of Computing.)

these complaints were often disguised versions of the solipsistic 'Argument from Consciousness': perhaps you could build a computer that *expressed* enjoyment of strawberries and cream, but would it really *feel* the pleasure? 'The criticism that a machine cannot have much diversity of behaviour is just a way of saying that it cannot have much storage capacity', which in the future, Turing implied, will not be true. He was right: desktop PCs have millions of times more memory than the Blue Pig, and have indeed many diverse applications. But there are in fact real teeth to this objection. As

Artificial Intelligence research attracted massive funding in the 1950s, its proponents predicted great advances. They soon found that some sorts of human behaviour were easy to imitate, but others were far, far more difficult than ever expected. Proficiency in natural language (i.e. things like spoken English – although automatic translation of Russian was the great Cold War goal) has broken each new generation of Artificial Intelligence machine. Each time the mechanical approach has failed, a choice has been offered: give up the mechanical dream or conclude that language is more complicated than expected and ask for better machines. So far the latter choice has won out.

In general, Turing argued that there was no reason to think that machines should not be able to think. It was a deeply materialist view – appeals to souls or other insubstantial essences that might make humans different were given short shrift. Twentieth-century theologians would attack the 'God of the gaps', arguing that a defence of religion that depended on invoking the deity only where there was ignorance was a weak and dangerous position. Analogously, Turing attacked the 'Mind of the gaps':

The 'skin of an onion' analogy is . . . helpful. In considering functions of the mind or the brain we find certain operations which we can explain in purely mechanical terms. This we say does not correspond to the real mind: it is a sort of skin which we must strip off if we are to find the real mind. But then in what remains we find a further skin to be stripped off, and so on. Proceeding in this way do we ever come to the 'real' mind, or do we eventually come to the skin which has nothing in it? In the latter case the whole mind is mechanical.

Here was a research programme, indeed one that had already begun. In 1943, a collaborative paper written in Chicago by Warren McCulloch and the young, brilliant but self-destructively unstable Walter Pitts had proposed that neurons, the basic units of the brain, acted like logic circuits. (McCulloch later claimed that this insight was inspired by Turing's paper on computable numbers.) Here was one of the inner skins of the onion already reduced to mechanism. A galaxy of interconnected, if occasionally squabbling, new scientific specialities sought to illuminate this new research programme: cybernetics, Artificial Intelligence and cognitive psychology, to name just three.

In the United States, the programme soon dominated psychology, the science of the mind. In the 1930s, the guiding paradigm had been Behaviourism, which insisted that only the study of the external actions of humans and animals could be the basis of proper science, and internal phenomena like mind were treated like a black box, only to be judged by observable external reactions to stimuli. Inspired by the spread of the universal machine, the new cognitive psychology of the 1950s opened the black box and declared it to be a computer. The brain and the mind were cast as aspects of an information-processing machine.

So do we have here a clue to the success of the universal machine? Was the machine in fact so universal that in 1948 there were already several billion examples walking the earth? If the brain was a computer, was it a surprise that a computer could replace or imitate humans?

Over the past 500 years, various mechanical models have been used to account for the inner life of men and women. In early modern Europe, the device upon which the model was based might have been a musical instrument, such as a lute: a note from a struck string caused a similar string to resonate. In this way, emotions were shared (think of the phrase 'she plucked my heartstrings'). In the eighteenth century, natural philosophers appealed to clockwork or balances, and adopted Newton's mechanical language of forces. In Freudian psychoanalysis, the imagery was a mix of hydraulics and thermodynamics: unconscious pressures and sexual energies, or the human being as steam engine. Explanations of mental life have often been made by appealing to a machine favoured in that period. Perhaps we should now ask, not whether the mind is in fact a computer, but what was special about the twentieth century that convinced so many that the computer was a good model for the mind.

20   Cold War Minds

I have argued that the universalism of the computer was partly an appeal to the universalism of mind. This appeal had old roots in Enlightenment philosophy – the idea that all humans possessed reason, and that the application of reason would inevitably lead to progress. In the 1950s, Noam Chomsky, then a young professor at MIT, had sought to demolish behaviourism by postulating 'universal' innate grammatical structures – a hard-wired ability to learn language. This sounds like Enlightenment rationalism – the belief in universal reason – but it resonated because of where he was speaking from. To see this point, we need to look at how and why the computer industry rapidly spread.

In the post-war decades, computers went from experimental scientific machines to the foundation of a billion-dollar industry. In the early 1950s, production runs of commercial computers numbered in the tens. By the late 1950s, sales of IBM's 1401 machine ran into the tens of thousands. In 1964, IBM launched a new compatible series of machine, ranging from the supercomputer to the small office computer, a strategy that was risky but profitable, generating that company alone a revenue of over $5 billion a year. IBM's success stemmed partly from a superb sales force, but also partly because it was a major beneficiary of Cold War largesse. Funding from the United States military not only helped the company through the difficult early years of electronic computer development, but also guaranteed a market for IBM products, and made accessible to the company military research findings. The United States was willing to support private industry to this unprecedented degree because of the ideological struggle with the other global superpower, the Soviet Union. Cold War funding put research money into the pockets of computer scientists, and gave them the best laboratories in the world – at MIT, Stanford University, IBM – to work in. The Cold War produced computers, while the computer also became a guiding metaphor for the Cold War.

To give one example: the United States Navy and

Air Force paid for the development of the prototype Whirlwind computer at MIT in 1951. Whirlwind was to form the centre of the United States' real-time command and control system, SAGE, that collected together the data from radar stations surrounding the country, displayed incoming bombers, and controlled the response. SAGE was an iconic Cold War technological system, symbolising what the historian Paul Edwards calls a 'closed world'. In developing Whirlwind, the scientists at MIT made several innovations: working out how to display digital information graphically on video displays; designing larger memories to hold the data on the missiles and aircraft; and computer languages to tell the system what to do. IBM built 56 computers for SAGE, receiving $30 million for each – as well as knowledge of state-of-the-art techniques.

The Cold War also shaped the development of computing in Britain. The collaboration between Manchester University and Ferranti, which led to the first commercial electronic computers, was enabled by funding from the Ministry of Supply, the government department responsible for overseeing defence research, including both the guided and atomic weapons programmes, which were voracious consumers of computing power. Military funding in Britain and the United States pushed the development of computing throughout the Cold War period, resulting in milestones such as airborne computers (in the aborted British TSR-2 aircraft) and the ARPANET (the origin of the Internet).

But beneficiaries of projects such as SAGE also included psychology. How did the human operator, absorbing the information displayed on the radar screen, fit into the system? The answer was simple: with the rest of the machine understood in terms of information processing, the human mind was recast along similar lines. For much of the 1950s and 1960s, studies of the mind were tied, either directly or indirectly, to this dominant research programme, which had its centres in laboratories such as those at MIT. Cognitive psychology – with its apparently *universal* claims that the human mind acts like a computer – had very *specific* roots in well-funded Cold War projects such as SAGE.

One man, however, was a victim rather than a beneficiary of the Cold War climate. In 1954, Turing probably committed suicide, dying after biting an apple laced with potassium cyanide. There was no necessary causal link between the trauma of his prosecution for homosexuality two years previously, after which he had been placed on compulsory hormone treatment, and the mathematician's death. The year 1952 had been bleak, but 1954 was a happier period. However, Cold War paranoia meant that Turing's private life was increasingly the concern of the British secret service. A consultancy role with GCHQ was abruptly severed in 1950. When Turing visited Norway in 1952, the intelligence service must have had fits: he knew the secrets of code-breaking; he was an expert in a radical new technology of military application; Norway shared a border with the Soviet Union; and he was, in their eyes, open to blackmail. Thus Hodges describes this increasingly hostile atmosphere – in which one of the key figures of Bletchley Park was cast as a potential traitor. Hugh Whitemore, in the play *Breaking the Code*, drops a hint of threat in the words of 'John Smith', the man from secret Whitehall: 'We have to make sure this knowledge is protected.'

21   Materialisation

We have seen that the spur of war has played an important role in forcing the pace of technological change, privileging speed of operation and decision making. The *global* conflicts of the twentieth century therefore form part of the explanation of the universal machine – from World War to World Wide Web, if you like. But they are not quite the whole story. I have stressed that a long history, including Ancient Greek mathematics and eighteenth-century philosophy, but especially emphasising nineteenth-century industrialisation and twentieth-century warfare, is needed to understand the computer. The remaining brick to be put in place concerns decision making more broadly, a matter of changing government and business.

Let us rewind back to Charles Babbage. Turing described Babbage's Analytical Engine as a 'universal digital computer', but Babbage himself put a revealing spin on what he understood his Engine to be. In his autobiography, he recalls, in a passage worth quoting at length:

In 1840 I received from my friend M. Plana a letter pressing me strongly to visit Turin at the then approaching meeting of Italian philosophers. In that letter M. Plana stated that he had enquired anxiously of many of my countrymen about the power and mechanism of the Analytical Engine. He remarked that from all the information he could collect the case seemed to stand thus:

'Hitherto the legislative department of our analysis has been all powerful – the executive all feeble.

Your engine seems to give us the same control over the executive which we have hitherto only possessed over the legislative department.'

Considering the exceedingly limited information which . . . could have reached my friend respecting the Analytical Engine, I was equally surprised and delighted at his exact provision of its powers. Even at the present moment [1864]

I could not express more clearly, and in fewer terms its real object.

In Babbage's own best summary of the 'real object' of the Analytical Engine, therefore, it was to be a *political* machine, and one in which automatic control would be gained over the entirety of the political process, both decision making and administrative. (Babbage's revolutionary bid for control over the executive – or at least control by proxy of the machine – was a radical solution, to say the least, to his frustrations of securing funding for his Engines.)

Was this mere analogy, or was there a deeper point being made? I think the latter. Again, we must consider the broader context. The role of government was changing in the nineteenth century. Not only (in Britain, say) was electoral reform opening up control of the state to the middle classes, but the first glimmerings of the welfare state were showing. Over the next century, more and more new roles would become the responsibility of government (public health, education, insurance, pensions, and so on). The state grew, and so did the number of state employees. By the end of the nineteenth century, the civil service employed tens of thousands. Furthermore, how was this executive part of government to be trusted, since it was now not only large but also employed members of the lower classes (and even women) that could not automatically be trusted by the gentlemanly elite? The solution was simply, but powerfully, a matter of language: cast the civil service as a machine. A civil service 'machine' would be neutral, interest-free, even efficient, and applicable to any task. (Why this worked is a story of educational reform, liberal politics and specific struggles over the nature of the nineteenth-century state, which is interesting but not crucial.) By 1900, if you searched Britain for a general-purpose machine of universal application, you would be led to the civil service.

Of course, the civil servants themselves were human. But they did come in two kinds. Following reform initiated in 1854, the British civil service was split into two: the generalist 'intellectuals' and the rule-following 'mechanicals' (with the *whole* still cast as a machine). This split has survived to the present day, and was replicated elsewhere, notably early on in the Indian Civil Service, a frequent laboratory of administrative experiment. We should immediately notice two things: first, that the separation of generalist from rule-following clerk precisely mirrored Babbage's separation of instructions and mill in the Analytical Engine; second, that the paradox of the computer, that it is something of infinite application, yet all it does is follow instructions, is also present in the combination of the generalist and the 'mechanical' clerk.

Something similar was occurring in nineteenth-century business, especially in the United States. It has already been argued that as the railroads snaked across the geographical immensity of the new world, they were hit by crises of control. The response was the creation of vertical hierarchies of managers who sought to control their dispersed business empires through information and communication. The similarities between the American case (where business was relatively more important than the state) and the British (where the opposite was true) lie in the creation and formalisation of managerial hierarchies, and a parallel story of mechanisation as Hollerith's system spread through both countries in the late nineteenth and early twentieth centuries. The difference lies in universality. Put crudely, business corporations had narrow aims – rarely more than the monopolisation of an industrial sector – whereas those of governments were general, universal.

There is a third perspective to Turing's astounding 1937 paper that I have not discussed yet (the other two being the solution of Hilbert's *Entscheidungsproblem* and, in doing so, imagining the universal computing machine – it also suggested, in passing, the computational model of the mind). When he considered his human computer at work, he argued that it could be captured entirely – 'determined' – by the human computer's 'state of mind' and the symbols written on the paper tape on the desk. Significantly, this, Turing argued, can proceed in two ways. In the first, the flexibility of the 'state of mind' was all important; in the second, the same work was transferred to rule-following:

It is always possible for the computer to break off from his work, to go away and forget all about it, and later to come back and go on with it. If he does this he must leave a note of instructions (written in some standard form) explaining how the work is to be continued.

This is a picture of the generalist popping out to lunch while the mechanical clerk dependably continues. Turing's universal machine, by extension, contained the generalist–mechanical split that defined that other universal executive machine of state. (Turing, recall, was steeped in the family culture of the Indian Civil Service through his father.)

At Bletchley Park, if you had the security clearance, you could read in the daily Ultra summaries clear evidence of the profound challenges faced by states as they struggled in the deepest of conflicts. The code-breaking centre is a key site for understanding twentieth-century history, not because of the gathering of geniuses there, nor even because the knowledge it produced shaped the conduct of the war. Instead, its greatest significance is as an industrialised bureaucracy hit by crises of both industrial and bureaucratic control. Bletchley Park pointed to the future in its almost entire preoccupation with the fast manipulation of symbolic information. In a pattern that can be read repeatedly in the last two hundred years, crises of control were answered by innovations in information technologies: the railroad crashes and the telegraph;

cholera epidemics and public health statistics; or the United States census and the Hollerith system. Bletchley Park – as a manifestation of civil service in wartime – materialised Turing's universal machine.

In perhaps the most thoughtful reflection on the relationships between the universal machine and society, Joseph Weizenbaum, in his *Computer Power and Human Reason*, summarised his views in a way that will be familiar to many. It is contained in a discussion of the compulsive programmer. These characters are

bright young men of dishevelled appearance, often with sunken glowing eyes [and] *can be seen sitting at computer consoles, their arms tensed and waiting to fire their fingers, already poised to strike, at the buttons and keys on which their attention seems to be riveted as a gambler's on the rolling dice . . . Their rumpled clothes, their unwashed and unshaven faces, and their uncombed hair all testify that they are oblivious to their bodies and to the world in which they move. They exist, at least when so engaged, only through and for the computers.*

The universal machine has given the compulsive programmer an unrestricted choice of laws, and omnipotence within a micro-world. Weizenbaum echoes a widespread fear:

> [W]e can take a continuum. At one of its extremes stand scientists and technologists who much resemble the compulsive programmer. At the other extreme are those scientists, humanists, philosophers, artists, and religionists who seek understanding as whole persons and from all possible perspectives. The affairs of the world appear to be in the hands of technicians whose psychic constitutions approximate those of the former to a dangerous degree.

So, in the hands of the compulsive programmers, the world is becoming more like a computer. I think the reverse is more historically accurate. As a materialisation of bureaucracy and managerial capitalism, the universal machine was made like the world.

Figure 10: Computers as a materialisation of bureaucracy. Why did computers fit so well into the big managerial corporations and public government departments of the twentieth century? Because computers were made in their image. The photograph shows System 4 and ICL 1901 computers at the Department of Health and Social Security, *c.* 1970. (Source: National Archive for the History of Computing.)

Further Reading

Agar, Jon. *The Government Machine*. Cambridge, Massachusetts: The MIT Press, 2002.
The universal machine as a materialisation of bureaucracy.

Babbage, Charles. *Passages in the Life of a Philosopher*. 1864. Modern edition edited by Martin Campbell-Kelly. London: William Pickering, 1994.
A great read, with descriptions of both Difference and Analytical Engines, as well as of cantankerous campaigns against 'street nuisances'.

Beniger, James. *The Control Revolution: Technological and Economic Origins of the Information Society*. Cambridge, Massachusetts: Harvard University Press, 1986.
The proponent of the idea that information technology innovations follow crises of control stemming from the industrial revolution.

Campbell-Kelly, Martin, and Aspray, William. *Computer: A History of the Information Machine*. New York: Basic Books, 1996.

Ceruzzi, Paul. *A History of Modern Computing*. Cambridge, Massachusetts, and London: MIT Press, 1998.
Both of these provide very good introductions to the history of computing.

Chandler, Alfred. *The Visible Hand: The Managerial Revolution in American Business*. Cambridge, Massachusetts: Belknap Press, 1977.
Classic of business history. Source of understanding the rise of management and the corporation.

Edwards, Paul N. *The Closed World*. Cambridge, Massachusetts: MIT Press, 1996.
Ground-breaking history of computing: the computer as Cold War icon and artefact.

Hodges, Andrew. *Alan Turing: The Enigma of Intelligence*. London: Burnett Books, 1983.
A biography that cannot be bettered. See also Hodges' website at www.turing.org.uk

Lubar, Steven. *Infoculture*. Boston and New York: Houghton Mifflin, 1993.
A cultural history of information-handling machines.

Smith, Roger. *Fontana History of the Human Sciences*. London: Fontana, 1997.

A new synthetic history – a good source on psychology, political economy and much else.

Weizenbaum, Joseph. *Computer Power and Human Reason*. London: Freeman, 1976.

The creator of the computer-as-psychiatrist program, ELIZA, provides profound insights into the spread and impact of computers.

Williams, Michael R. *A History of Computing Technology*. Englewood Cliffs, New Jersey: Prentice-Hall, 1985.

Good history, with emphasis on hardware.

Whitemore, Hugh. *Breaking the Code*. Oxford: Amber Lane Press, 1987.

A play based on Hodges' biography. First performed at the Arnauld Theatre, Guildford, in 1986, before moving later that year to the Theatre Royal, London.